BASIC LIMNOLOGY AND FISH BIODIVERSITY

BASIC LIMNOLOGY AND FISH BIODIVERSITY

Dr. Shalini Verma

RANDOM PUBLICATIONS

NEW DELHI - 110 002 (INDIA)

Basic Limnology and Fish Biodiversity

ISBN 978-93-51117-03-2

Published in 2015 in India by

RANDOM PUBLICATIONS

4376-A/4B, Gali Murari Lal, Ansari Road
New Delhi-110 002
Phone: +9111-43580356, 23289044
E-mail: randomexports@gmail.com; sales@randompublications.com; info@randompublications.com

Type Setting by: Friends Media, Delhi-110089
Printed at Thomson Press(India) Limited

Preface

Limnology, also called freshwater science, is the study of inland waters. It is often regarded as a division of ecology or environmental science. It covers the biological, chemical, physical, geological, and other attributes of all inland waters (running and standing waters, both fresh and saline, natural or man-made). This includes the study of lakes and ponds, rivers, springs, streams and wetlands. A more recent sub-discipline of limnology, termed landscape limnology, studies, manages, and conserves these aquatic ecosystems using a landscape perspective.

Limnology is closely related to aquatic ecology and hydrobiology, which study aquatic organisms in particular regard to their hydrological environment. Fish rule the oceans and freshwater of the world. They are the most common group of vertebrates (creatures with backbones), and were also the first vertebrates. Most species only live in fresh or salt water, but some that live in the ocean come into freshwater to spawn, and a few can live in either fresh or salt water. Although most fish species have probably been discovered and described, about 250 new ones are still discovered every year. According to FishBase, 32,800 species of fish had been described by June 2014. Coral reefs in the Indo-Pacific constitute the centre of diversity for marine fishes, whereas continental freshwater fishes are most diverse in large river basins of tropical rainforests, especially the Amazon, Congo, and Mekong basins. More than 5,600 fish species inhabit Neotropical freshwaters alone, such that Neotropical fishes represent about 10% of all vertebrate species on the Earth. Freshwater fish are those that spend some or all of their lives in fresh water, such as rivers and lakes, with a salinity of less than 0.05%. These environments differ from marine conditions in many ways, the most obvious being the difference in levels of salinity. To survive fresh water, the fish need a range of physiological adaptations. 41.24% of all known species of fish are found in fresh water. This is primarily due to the rapid speciation that the scattered habitats make possible. When dealing with ponds and lakes, one might use the same basic models of speciation as when studying island biogeography. Freshwater fish differ physiologically from salt water fish in several respects. Their gills must be able to diffuse dissolved gasses while keeping the salts in the body fluids inside. Their scales reduce water diffusion through the skin: freshwater fish that have lost too many scales will die. They also have well developed kidneys to reclaim salts from body fluids before excretion. India is rich with inland freshwater fish, with about 940 species known from its rivers, lakes and estuaries. This constitutes about 38% of the Indian Ichthyofauna and are of considerable economic and scientific value. Of these about 500 species are primary freshwater fish with around 65% endemic, cloistered in the two hot spots of India, the Western Ghats and the North East. Many of these are unique to certain stretches of the various rivers especially the upper reaches and many more new species are being reported from these forested hills. However, threats to these fauna are aplenty, with urbanization, deforestation, habitat loss, pollution, over-harvesting, and culture of exotics. The only way for a fish enthusiast to save the fish and its habitats is creating awareness among the people. In this context, knowledge of the fish fauna of a region is the first step towards conservation.

Written in an engaging and accessible style, the book considers the value of preserving aquatic biodiversity offers an overview of imperiled fishes on a taxonomic and geographic basis presents a synthesis of common characteristics. This is a comprehensive book on the biodiversity of one of the most diverse ecosystems known - tropical freshwater.

I thank all members of my team who have helped in the preparation of the book. My special thanks go to "Random Publications" who have published the book.

— *Dr. Shalini Verma*

Contents

Chapter 1

Basic Limnology

Limnology, also called freshwater science, is the study of inland waters. It is often regarded as a division of ecology or environmental science. It covers the biological, chemical, physical, geological, and other attributes of all inland waters (running and standing waters, both fresh and saline, natural or man-made). This includes the study of lakes and ponds, rivers, springs, streams and wetlands. A more recent sub-discipline of limnology, termed landscape limnology, studies, manages, and conserves these aquatic ecosystems using a landscape perspective.

Limnology is closely related to aquatic ecology and hydrobiology, which study aquatic organisms in particular regard to their hydrological environment.

The term limnology was coined by François-Alphonse Forel (1841–1912) who established the field with his studies of Lake Geneva. Interest in the discipline rapidly expanded, and in 1922 August Thienemann (a German zoologist) and Einar Naumann (a Swedish botanist) co-founded the International Society of Limnology (SIL, for originally Societas Internationalis Limnologiae). Forel's original definition of limnology, "the oceanography of lakes", was expanded to encompass the study of all inland waters, and influenced Benedykt Dybowski's (Polish naturalist and physician) work on Lake Baikal.

Landscape limnology is the spatially explicit study of lakes, streams, and wetlands as they interact with the freshwater, terrestrial, and human landscapes to determine the effects of pattern on ecosystem processes across temporal and spatial scales. Limnology is the study of inland water bodies inclusive of rivers, lakes, and wetlands; landscape limnology seeks to integrate all of these ecosystem types.

The terrestrial component represents spatial hierarchies of landscape features that influence what materials, whether solutes or organisms, are transported to aquatic systems; aquatic connections represent how these materials are transported; and human activities reflect features that influence both what and how these materials are transported as well as their quantity and temporal dynamics.

The core principles or themes of landscape ecology provide the foundation for landscape limnology. These ideas can be synthesized into a set of four landscape ecology themes that are broadly applicable to any aquatic ecosystem type, and that consider the unique features of such ecosystems.

A landscape limnology framework begins with the premise of Wiens (2002): freshwater ecosystems can be considered patches. As such, the location of these patches and their placement relative to other elements of the landscape is important to the ecosystems and their processes. Therefore, the four main themes of landscape limnology are:

(1) Patch characteristics: The characteristics of a freshwater ecosystem include its physical morphometry, chemical, and biological features, as well as its boundaries. These boundaries are often more easily defined for aquatic ecosystems than for terrestrial ecosystems (e.g., shoreline, riparian zones, and emergent vegetation zone) and are often a focal-point for important ecosystem processes linking terrestrial and aquatic components.

(2) Patch context: The freshwater ecosystem is embedded in a complex terrestrial mosaic (e.g., soils, geology, and land use/cover) that has been shown to drive many within-ecosystem features and processes such as water chemistry, species richness, and primary and secondary productivity.

(3) Patch connectivity and directionality: The complex freshwater mosaic is connected to the particular patch of interest and defines the degree to which materials and organisms move across the landscape through freshwater connections. For freshwater ecosystems, these connections often display a strong directionality component that must be explicitly considered. For example, a specific wetland can be connected through groundwater to other wetlands or lakes, or through surface water connections directly to lakes and rivers, or both, and the directionality of those connections will strongly impact the movement of nutrients and biota.

(4) Spatial scale and hierarchy: Interactions among terrestrial and freshwater elements occur at multiple spatial scales that must be considered hierarchically. The explicit integration of hierarchy into landscape limnology is important because (a) many freshwater ecosystems are hierarchically organized and controlled by processes that are hierarchically organized, (b) most freshwater ecosystems are managed at multiple spatial scales, from policy set at the national level, to land management conducted at local scales, and (c) the degree of homogeneity among freshwater ecosystems can change in relation to the scale of observation.

Contributions to Other Fields

Findings from landscape limnology research are contributing to many facets of aquatic ecosystem research, management, and conservation. Landscape limnology is especially relevant for geographical areas with thousands of ecosystems (i.e. lake-rich regions of the world), in situations with a range of human disturbances, or when considering lakes, streams, and wetlands that are connected to other such ecosystems. For example, landscape limnology perspectives have contributed to the development of nutrient criteria for lakes, formation of classification systems that can be used to monitor the health of aquatic ecosystems, understanding ecosystem responses to environmental stressors, or explaining biogeographic patterns of community composition.

Historical Development and Scope of Limnology

The origins of limnology date back many centuries to a time when scientists were called natural philosophers and science was explored by a few, usually wealthy, individuals. For example, Gorham (1953) traced the development of wetland ecology and some of its fundamental premises to studies of British natural philosophers going back to the fifteenth and sixteenth centuries. Similarly, Hutchinson (1967) traced studies on lakes at least as far back as a fifteenth century study on the ponds of a European abbey. However, the evolution of limnology into a modern science depended on the development of concepts and tools in biology, chemistry, and physics that were not available until the late nineteenth century.

Although modern limnology encompasses the study of all inland waters, its development is particularly identified with the study of lakes. Much of the conceptual framework around which the science was built was derived from studies on lakes, and most of the early limnologists were lake scientists. A notable exception was Stephen Forbes, a principal architect of the framework for limnology who was trained as a fish biologist and spent most of his career working on rivers and streams in Illinois, a state with relatively few natural lakes. It is understandable why early limnologists focused on lakes. As systems with easily recognizable boundaries and long residence times for water and substances in it, they are more obvious subjects for systematic scientific analysis than are the open, flowing waters of streams and spatially less defined wetlands. Nonetheless, it must be noted that the historical treatment usually accorded to limnology is coloured by the fact that it generally is written by lake scientists,

The beginnings of limnology as a modern science usually are traced to the work of a few late nineteenth century biologists who focused on lake studies. The founders of lacustrine limnology defined the scope and nature of the field in a way that survives remarkably intact to the present day; they viewed the subject broadly and integratively.

While many believe Peter Erasmus Muller is credited with starting the foundations of limnological research by his plankton discoveries in fresh water, and while it appears that Anton Fritsch began investigations in the Bohemian Forest as early as 1871, it is still F. A. Forel (1841-1912) who is credited as the founder of limnological studies.

Francois Forel was the first scientist to use the term limnology in a publication. His three-volume treatise on Lake Geneva (bordered by Switzerland and France), published over the period 1892 to 1904, is considered the first book on limnology, and it was encyclopedic in scope. Its 14 chapters define the main supporting fields of modern lake limnology and reinforce the idea that lake limnology is the application of all relevant basic sciences to the analysis of lakes as fundamental units of study.

The integrative nature of limnology was stressed even before Forel coined the term limnology. In a prescient article published in 1887, Stephen Forbes described lakes as "microcosms," or little worlds. Although the term "ecosystem" was not introduced for another half century, Forbes defined an approach that presaged this concept. He proposed that lake studies should focus on many of the processes that today define the field of ecosystem ecology: mineral cycling, production and decomposition of organic matter, food web interactions and their impacts on the structure of biological communities, and the effects of physical conditions on biological communities. Forbes viewed these topics as essential to understanding lakes as functioning, integrated systems. The notion of lakes as microcosms (or integrated ecosystems) has pervaded their study ever since Forbes' time, even though the concept has Francois A. Forel invented the word limnology. The science would have been called "limnography," to match its sister science oceanography, had it not been for the priority of reserving the term "limnograph" for a device used to measure water height in lakes.

Forel was born in Morzes on the shore of Lake Geneva (known to the Swiss as Lac Leman). When he was 13, his father introduced him to "... the art of observing and questioning nature," according to a monograph that he later wrote about Lake Geneva. After graduating from the Academie de Geneve, Forbes completed a medical degree at Würzburg and taught there for three years. In 1870, he joined the faculty of the Academie de Lausanne, where he taught anatomy and physiology and started a lifelong study of Lake Geneva.

One of Forel's earliest observations connected the physical, chemical, and biological properties of Lake Geneva. While looking to see if waves had left ripple marks on the bottom along the shore, he noticed a wriggling nematode in a sample of mud. This "poor worm" piqued his curiosity and led him to invent a bottom dredge with which he discovered that the depths of the lake were not a desert but were occupied by a specialized fauna rich in species and individuals. His studies resulted in a long series of influential papers on benthic fauna, their environmental conditions, and their significance to the fish population; the papers were consistent with modern concepts of ecosystem ecology. He followed with comparative work on other Swiss lakes.

In physical limnology, much of Forel's attention focused on water oscillations that create standing wave patterns, known to limnologists as seiches. In addition to obtaining massive data on Lake Geneva itself, he studied movements of water in small, tilted model lake basins, thus anticipating by many years a kind of experimental limnology. From these studies he developed generalizations about the relations between the dimensions of lakes and the periodicity of their seiches. He also conducted pioneering work in other aspects of physical limnology, devising a colour scale and studying light penetration with photographic paper, and considered external influences on the lake, particularly the relation of the water supply to glaciers.

Henri LeBlanc (1912) listed Forel's publications in categories: limnology (126 titles), glaciology (66), seismology (12), meteorology (19), natural history (28), archaeology (12), history (10), and biographies (15). The massive Lake Geneva monograph *Le Leman: Monographie*

Limnologique was published in three volumes in 1882, 1895, and 1904, with a total of 14 chapters. A small textbook appeared in 1901. He continued to work until a few months before his death in 1912, producing about 35 percent of his publications after the appearance of the last volume of the monograph in 1904.

While in the USA, another early pioneer that should take some of the credit is Stephen Alfred Forbes (1844-1930). Forbes was the first Chief of the Illinois Natural History Survey, a founder of aquatic ecosystem science and a dominant figure in the rise of American ecology. His publications are influential for their extensive field observations with conceptual insights. Forbes believed that ecological knowledge was fundamental for human well-being. While already famous as an economic entomologist, Forbes undertook studies of massive fish mortality in Lake Mendota, Wisconsin. He showed the connection of algae blooms and lake physics to fish kills, and embarked on a remarkable research programme into the ecology of lakes and rivers. Many of his insights about lake ecology were collected in an influential paper, "The Lake as a Microcosm" (Forbes, 1887). Forbes' work foreshadowed the ecosystem concept as well as modern ideas of behavioural ecology and food web dynamics. However, it should be mentioned that before Forbes had published any papers a few important advancements had been made. In 1865, Pietro Angelo Secchi (1818-1878) created the Secchi disk to measure water transparency, a very low tech piece of equipment which is still used today. Lagos Winkler in 1888 developed a field test for ascertaining the amount of dissolved oxygen in a sample of water.

During these early days of Limnology, Forel and many other scientists were only describing the physical, chemical, and biological aspects of a single lake. It was not until 1905 when C Wesenberg-Lund released a paper that changed the way limnologists studied the inland water systems. His paper was a comparative study of lakes in Scotland and Denmark, as well as some pioneering work on plankton and littoral animals. But despite some major advancement in the field, World War 1 broke out and effectively bought an end to this era.

Post World War, a new generation of Limnologists in Europe, USA, USSR and later in Japan led the advancements in the field of Limnology. In Europe there were two main individuals that made great strides to making up for the lost years while war raged across Europe. They were August Friedrich Thienemann (1882-1960) and Einar Christian Leonard Naumann (1891–1934), who in 1922 formed the International Society of Limnology (SIL). The aim of SIL was (and still is) to study and understand all aspects of limnology. The main way they achieved this was by the organization of triennial congresses, to promote scientific collaboration among those pursuing "real" research, and those who were more concerned with practical fishery, pollution, and water-supply problems.

August Thienemann was an ecologist, whose area of study was around individual insects but he was also one of the pioneers in the area of functional groupings. This lead Limnology towards ecological understandings of aquatic systems. His study of insects was used to try and determine the aquatic ecosystem based on the chironomid found. He focused on the society of organisms. Strom (1929) agrees with this focus, but goes on to say that the study should be extended further into the investigation of the single organism, including morphology, physiology and genetics.

Lindeman (1942) released an paper based on Thienemann's ideas, and the idea of nutrient cycling within lakes and food web structure, showing that the idea put forward by Thienemann in 1925 has withstood the test of time. Reynolds (1999) suggests food web theory in the community structure found within a ecosystem, continued to be a dominant concern as late as the 1980/90's. While Thienemann and a great number of other scientists believed that lakes were closed systems, Naumann (1891–1934) saw them as open systems. His intuition was that the input of phosphorus, nitrogen and calcium into the system, had a direct effect on the amount and composition of phytoplankton. There has since been hundreds of papers published on the phosphorus relationship to lake function and structure.

Naumann believed that phytoplankton is an indicator of nutrient conditions, however technology was not available to directly test this theory. This led him to classify lakes based on their nutrient levels: oligotrophic (poorly nourished), eutrophic (well nourished) and later added mesotrohic (medium nourished). These terms are still used today. Naumann's classification was accepted by Thienemann, however Thienemann modified it to include dystrophic (defectively nourished) to account for the lakes high in coloured organic matter. This fourth group is solely based on water colour as opposed to nutrient content. Because Naumann was alone in thinking that lakes were open systems he is not accredited with as much influence as Thienemann in the study of Limnology for this era, and this meant lakes were studied as closed units of organizations well into the 1960's and maybe even to some extent till today.

In the USA there were two limnologists that influenced the direction Limnology was heading, and this was in a different direction to their colleagues in Europe (Kalff, 2002). Edward Birge (1851-1950) and Chancey Juday (1871-1944) from their Wisconsin School were moving away from the simplex notion of Limnology to more order and logic. The Wisconsin Natural History Survey paid for the comparative study of hundreds of lakes, while their university connections allowed them to call on chemists, biologists, bacteriologists and instrument makers, to new levels. With this group of people they made many new instruments and developed many new techniques. Their access to many lakes of different size and morphometry that lay in different basins, led them to dispute any possibility of generalization on a narrow band of lakes as was being carried out in Europe by Thienemann.

Unlike in Europe, Birge and Juday were defining many different classifications of lakes such as allotrophic and autotrophic, as well as other concepts regarding how lakes retrieved nutrients by different processes depending on in and out flows. Birge and Juday were also developing more quantitative comparative methods than their European coworkers, however they did not have even the most simple techniques for seeing the statistical difference. Another major step Birge and Juday bought to Limnology, in 1925, was the lake cauterization of having phosphorus present which to this day is a principal proxy used to predict lake functioning. This led them to setup the Trout Lake Station, funded by the University of Wisconsin.

Birge believed that the only real way to study lakes was from the field, meaning out on the water observing in its natural habitat, not from a laboratory (Spoolman, 2003). The trout lake station allowed a group of scientists to live and work in that environment. Birge hired Juday to lead the field team, while he took care of the fundraising side. Spurred on by Forbes' 1887 paper "The Lake as a Microcosm" he believed that in order to understand the parts and processes of a lake, you have to collect enough data over a longer period of time. 1911-1916 saw Birge and Juday and a team of graduates undertake filtering of some 2000 tons of water and 200 tons of foodstuffs of fish. From these findings he is quoted as saying "Here we had for the first time a definite notion of the total quantity of food and eaters handled by the lakes in the process of their housekeeping.... After nearly 20 years of experience in the study of lakes we had learned to ask the question: how does a lake keep house This was a defining moment for Birge and many of the people who followed him.

Limnology changed again with the work done by George Evelyn Hutchinson (1903–1991). Hutchinson was a product of a more modern era of science and could see the need to bring mathematics into the equation. He is best known for his studies into lake metabolism, biogeochemistry, paleolimnology, lake classification and plankton diversity all published in his Treatise on Limnology in three volumes. Unlike his predecessors, he agreed with Naumann and did not see lakes as a single unit, instead he hoped to be able to describe each component of the complete ecosystem, building a whole from the individual parts.

The idea that biological interactions happening under the water line were the most important factors in determining the behaviour of lakes, fits well with the models put forward by earlier work. After his career as a researcher Hutchinson primarily turned to the development of ideas based around, determinants of community composition and the development of ecological theory based on evolutionary principles, put forward by Charles Darwin. The ideas that Hutchinson developed about lake interactions being linked to chemical/physical environment has formed the basis of many further studies.

Between 1939-1945, the Second World War raged across many parts of Europe, this effectively bought an end to an era of Limnology, and it was not till the mid to late 1950's that the science started again. With technological advancements made during the war, new instruments using isotopes were created and with the ever growing number of computers available for statistical tests, and easier international travel, Limnology grew rapidly. Between 1960 and 1980 the major changes of limnology happened. Until this time most of the study areas were based around improving water quality or dealing with waste water, and many projects were conducted by single person, or at best with help from a couple of students.

A new scheme was setup creating the International Biological Programme, where teams were formed in economically strong countries and were able to study whole lakes or streams. This ecosystem ecology led to new insights into relationships between organisms in the ecosystem, and with huge data sets this led to new quantitative models being developed to explain the results found. Reynolds sums up the history thus far as "a deepening appreciation of the constraints the aquatic environment imposes on the selection and survival of its inhabitants". The slowing down of new discoveries in the 1980 and 90's has had detrimental effect on limnology in terms of funding for further research. Despite now knowing more than ever, we are seeing a shift away from free study by scientists and more towards government pushed problem solving. Most papers on Limnology are pre World War II, and since then it is difficult to find any paper that has really defined real change or advancements in the field. Most advancement seems to have been made in technological ways, with more accurate measuring methods and computers to run huge data sets.

However G. Harris' paper aptly named This is not the end of limnology (or of science): the world may well be a lot simpler than we think has presented evidence that ecosystems (and freshwater ecosystems in particular) may well be a lot simpler than we think. Buried deep within a very complex world there are some general modes of behaviour, determined by fundamental principles, which impart certain kinds of high level order and predictability. Thus there is a need for further research into the field of Limnology.

Aquatic Ecosystem

An aquatic ecosystem is an ecosystem in a body of water. Communities of organisms that are dependent on each other and on their environment live in aquatic ecosystems. The two main types of aquatic ecosystems are marine ecosystems and freshwater ecosystems.

Freshwater ecosystems are a subset of Earth's aquatic ecosystems. They include lakes and ponds, rivers, streams, springs, and wetlands. They can be contrasted with marine ecosystems, which have a larger salt content. Freshwater habitats can be classified by different factors, including temperature, light penetration, and vegetation.

Freshwater ecosystems cover 0.80% of the Earth's surface and inhabit 0.009% of its total water. They generate nearly 3% of its net primary production. Freshwater ecosystems contain 41% of the world's known fish species.

Freshwater ecosystems can be divided into lentic ecosystems (still water) and lotic ecosystems (flowing water).

Limnology (and its branch freshwater biology) is a study about freshwater ecosystems. It is a part of hydrobiology.

Original efforts to understand and monitor freshwater ecosystems were spurred on by threats to human health (ex. Cholera outbreaks due to sewage contamination). Early monitoring focussed on chemical indicators, then bacteria, and finally algae, fungi and protozoa. A new type of monitoring involves differing groups of organisms (macroinvertebrates, macrophytes and fish) and the stream conditions associated with them.

Current biomonitering techniques focus mainly on community structure or biochemical oxygen demand. Responses are measured by behavioural changes, altered rates of growth, reproduction or mortality. Macroinvertebrates are most often used in these models because of well known taxonomy, ease of collection, sensitivity to a range of stressors, and their overall value to the ecosystem. Most of these measurements are difficult to extrapolate on a large scale, however.

The use of reference sites is common when assessing what a healthy freshwater ecosystem should "look like". Reference sites are easier to reconstruct in standing water than moving water. Preserved indicators such as diatom valves, macrophyte pollen, insect chitin and fish scales can be used to establish a reference ecosystem representative of a time before large scale human disturbance.

Common chemical stresses on freshwater ecosystem health include acidification, eutrophication and copper and pesticide contamination.

Over 123 freshwater fauna species have gone extinct in North America since 1900. Of North American freshwater species, an estimated 48.5% of mussels, 22.8% of gastropods, 32.7% of crayfishes, 25.9% of amphibians, and 21.3% of fishes are either endangered or threatened. Extinction rates of many species may increase severely into the next century because of invasive species, loss of keystone species, and species which are already functionally extinct. Projected extinction rates for freshwater animals are around five times greater than for land animals, and are comparable to the rates for rainforest communities. Recent extinction trends can be attributed largely to sedimentation, stream fragmentation, chemical and organic pollutants, dams, and invasive species.

Hydrobiology

Hydrobiology is the science of life and life processes in water. Much of modern hydrobiology can be viewed as a sub-discipline of ecology but the sphere of hydrobiology includes taxonomy, economic biology, industrial biology, morphology, physiology etc. The one distinguishing aspect is that all relate to aquatic organisms. Much work is closely related to limnology and can be divided into lotic system ecology (flowing waters) and lentic system ecology (still waters).

One of the significant areas of current research is eutrophication. Special attention is paid to biotic interactions in plankton assemblage including the microbial loop, the mechanism of influencing water blooms, phosphorus load and lake turnover. Another subject of research is the acidification of mountain lakes. Long-term studies are carried out on changes in the ionic composition of the water of rivers, lakes and reservoirs in connection with acid rain and fertilisation. One goal of current research is elucidation of the basic environmental functions of the ecosystem in reservoirs, which are important for water quality management and water supply.

Much of the early work of hydrobiologists concentrated on the biological processes utilised in sewage treatment and water purification especially slow sand filters. Other historically important work sought to provide biotic indices for classifying waters according to the biotic communities that they supported.

This work continues to this day in Europe in the development of classification tools for assessing water bodies for the EU water framework directive.

Fish Diversity

Fish are very diverse animals and can be categorised in many ways. This article is an overview of some of ways in which fish are categorised. Although most fish species have probably been discovered and described, about 250 new ones are still discovered every year. According to FishBase, 32,800 species of fish had been described by June 2014. That is more than the combined total of all other vertebrate species: mammals, amphibians, reptiles and birds.

Fish species diversity is roughly divided equally between marine (oceanic) and freshwater ecosystems. Coral reefs in the Indo-Pacific constitute the centre of diversity for marine fishes, whereas continental freshwater fishes are most diverse in large river basins of tropical rainforests, especially the Amazon, Congo, and Mekong basins. More than 5,600 fish species inhabit Neotropical freshwaters alone, such that Neotropical fishes represent about 10% of all vertebrate species on the Earth. Exceptionally rich sites in the Amazon basin, such as Cantão State Park, can contain more freshwater fish species than occur in all of Europe.

Fish systematics is the formal description and organisation of fish taxa into systems. It is complex and still evolving. Controversies over "arcane, but important, details of classification are still quietly raging."

The term "fish" describes any non-tetrapod chordate, (i.e., an animal with a backbone), that has gills throughout life and has limbs, if any, in the shape of fins. Unlike groupings such as birds or mammals, fish are not a single clade but a paraphyletic collection of taxa, including jawless, cartilaginous and skeletal types.

Jawless Fish

Jawless fish were the earliest fish to evolve. There is current debate over whether these are really fish at all. They have no jaw, no scales, no paired fins, and no bony skeleton. Their skin is smooth and soft to the touch, and they are very flexible. Instead of a jaw, they possess an oral sucker. They use this to fasten onto other fish, and then use their rasp-like teeth to grind through their host's skin into the viscera. Jawless fish inhabit both fresh and salt water environments. Some are anadromous, moving between both fresh and salt water habitats.

Extant jawless fish are either lamprey or hagfish. Juvenile lamprey feed by sucking up mud containing micro-organisms and organic debris. The lamprey has well-developed eyes, while the hagfish has only primitive eyespots. The hagfish coats itself and carcasses it finds with noxious slime to deter predators, and periodically ties itself into a knot to scrape the slime off. It is the only invertebrate fish and the only animal which has a skull but no vertebral column. It has four hearts, two brains, and a paddle-like tail.

Cartilaginous Fish

Cartilaginous fish have a cartilaginous skeleton. However, their ancestors were bony animals, and were the first fish to develop paired fins. Cartilaginous fish don't have swim bladders. Their skin is covered in placoid scales *(dermal denticles)* that are as rough as sandpaper. Because cartilaginous fish do not have bone marrow, the spleen and special tissue around the gonads produces red blood cells. Their tails can be asymmetric, with the upper lobe longer than the lower lobe. Some cartilaginous fishes possess an organ called Leydig's Organ which also produces red blood cells.

There are over 980 species of cartilaginous fish. They include sharks, rays and chimaera.

Bony Fish

Bony fish include the lobe finned fish and the ray finned fish. The lobe finned fish is the class of fleshy finned fishes, consisting of lungfish, and coelacanths. They are bony fish with fleshy, lobed paired fins, which are joined to the body by a single bone. These fins evolved into the legs of the first tetrapod land vertebrates, amphibians. Ray finned fishes are so-called because they possess lepidotrichia or "fin rays", their fins being webs of skin supported by bony or horny spines ("rays").

There are three types of ray finned fishes: the chondrosteans, holosteans, and teleosts. The chondrosteans and holosteans are among the earlier fish to evolve, and share characteristics with both teleosts and sharks. In comparison with the other chondrosteans, the holosteans are closer to the teleosts and further from sharks.

Teleosts

Teleosts are the most advanced or "modern" fishes. They are overwhelmingly the dominant class of fishes (or for that matter, vertebrates) with nearly 30,000 species, covering about 96 percent of all extant fish species. They are ubiquitous throughout fresh water and marine environments from the deep sea to the highest mountain streams. Included are nearly all the important commercial and recreational fishes.

Teleosts have a movable maxilla and premaxilla and corresponding modifications in the jaw musculature. These modifications make it possible for teleosts to protrude their jaws outwards from the mouth. The caudal fin is homocercal, meaning the upper and lower lobes are about equal in size. The spine ends at the caudal peduncle, distinguishing this group from those in which the spine extends into the upper lobe of the caudal fin.

Fish Diversity in Freshwater Systems

Although freshwater ecosystems such as rivers, lakes and wetlands occupy less than 2% of the earth's total land surface, they provide a wide range of habitats for a significant proportion of the world's plant and animal species. Many are yet to be discovered, but the number of freshwater species worldwide is estimated at between 9,000 and 25,000. However, this number is rapidly decreasing due to human interference.

Physical alteration, habitat degradation, excessive water withdrawal and pollution have contributed directly or indirectly to the decline in freshwater species. Other factors that reduce freshwater biodiversity include the incursion of non-native species and the mismanagement of inland fisheries. Today, an estimated 20% of the world's freshwater fish is vulnerable, endangered or extinct.

Freshwater Fishes of India

India is rich with inland freshwater fish, with about 940 species known from its rivers, lakes and estuaries. This constitutes about 38% of the Indian Ichthyofauna and are of considerable economic and scientific value. Of these about 500 species are primary freshwater fish with around 65% endemic, cloistered in the two hot spots of India, the Western Ghats and the North East. Many of these are unique to certain stretches of the various rivers especially the upper reaches and many more new species are being reported from these forested hills. However, threats to these fauna are aplenty, with urbanization, deforestation, habitat loss, pollution, over-harvesting, and culture of exotics. The only way for a fish enthusiast to save the fish and its habitats is creating awareness among the people. In this context, knowledge of the fish fauna of a region is the first step towards conservation.

Limnology of Rivers: The Lotic System

A river is a natural watercourse, usually freshwater, flowing towards an ocean, a lake, a sea, or another river. In some rare cases a river could flow into the ground and dry up completely at the end of its course, without reaching another body of water. Small rivers may be called by several other names, including stream, creek, brook, rivulet, and rill. There are no official definitions for generic terms, such as river, as applied to geographic features, although in some countries or communities a stream may be defined by its size. Many names for small rivers are specific to geographic location; examples are "run" in some parts of the United States, "burn" in Scotland and northeast England, and "beck" in northern England. Sometimes a river is defined as being larger than a creek, but not always: the language is vague.

Rivers are part of the hydrological cycle. Water generally collects in a river from precipitation through a drainage basin from surface runoff and other sources such as groundwater recharge, springs, and the release of stored water in natural ice and snowpacks (e.g. from glaciers). Potamology is the scientific study of rivers while limnology is the study of inland waters in general.

Extraterrestrial rivers have recently been found on Titan. Channels may indicate past rivers on other planets, specifically outflow channels on Mars and are theorised to exist on planets and moons in habitable zones of stars.

Topography

A river begins at a source (or more often several sources) and ends at a mouth, following a path called a course. The water in a river is usually confined to a channel, made up of a stream bed between banks. In larger rivers there is also a wider floodplain shaped by flood-waters over-topping the channel. Floodplains may be very wide in relation to the size of the river channel. This distinction between river channel and floodplain can be blurred, especially in urban areas where the floodplain of a river channel can become greatly developed by housing and industry.

Rivers can flow down mountains, through valleys (depressions) or along plains, and can create canyons or gorges.

The term upriver (or upstream) refers to the direction towards the source of the river, i.e. against the direction of flow. Likewise, the term downriver (or downstream) describes the direction towards the mouth of the river, in which the current flows.

The river channel typically contains a single stream of water, but some rivers flow as several interconnecting streams of water, producing a braided river. Extensive braided rivers are now found in only a few regions worldwide, such as the South Island of New Zealand. They also occur on peneplains and some of the larger river deltas. Anastamosing rivers are similar to braided rivers and are also quite rare. They have multiple sinuous channels carrying large volumes of sediment. There are rare cases of river bifurcation in which a river divides and the resultant flows ending in different seas. An example is the bifurcation of Nerodime River in Kosovo.

A river flowing in its channel is a source of energy which acts on the river channel to change its shape and form. In 1757, the German hydrologist Albert Brahms empirically observed that the submerged weight of objects that may be carried away by a river is proportional to the sixth power of the river flow speed. This formulation is also sometimes called Airy's law. Thus, if the speed of flow is doubled, the flow would dislodge objects with 64 times as much submerged weight. In mountainous torrential zones this can be seen as erosion channels through hard rocks and the creation of sands and gravels from the destruction of larger rocks. In U-shaped glaciated valleys, the subsequent river valley can often easily be identified by the V-shaped channel that it has carved. In the middle reaches where a river flows over flatter land, meanders may form through erosion of the river banks and deposition on the inside of bends. Sometimes the river will cut off a loop, shortening the channel and forming an oxbow lake or billabong. Rivers that carry large amounts of sediment may develop conspicuous deltas at their mouths. Rivers whose mouths are in saline tidal waters may form estuaries.

Throughout the course of the river, the total volume of water transported downstream will often be a combination of the free water flow together with a substantial volume flowing through sub-surface rocks and gravels that underlie the river and its floodplain (called the hyporheic zone). For many rivers in large valleys, this unseen component of flow may greatly exceed the visible flow.

River Ecosystem

The ecosystem of a river is the river viewed as a system operating in its natural environment, and includes biotic (living) interactions amongst plants, animals and micro-organisms, as well as abiotic (nonliving) physical and chemical interactions.

River ecosystems are prime examples of lotic ecosystems. *Lotic* refers to flowing water, from the Latin *lotus*, washed. Lotic waters range from springs only a few centimeters wide to major rivers kilometers in width. Much of this article applies to lotic ecosystems in general, including related lotic systems such as streams and springs. Lotic ecosystems can be contrasted with lentic ecosystems, which involve relatively still terrestrial waters such as lakes and ponds. Together, these two fields form the more general study area of freshwater or aquatic ecology.

The following unifying characteristics make the ecology of running waters unique from that of other aquatic habitats.

- Flow is unidirectional.
- There is a state of continuous physical change.
- There is a high degree of spatial and temporal heterogeneity at all scales (microhabitats).
- Variability between lotic systems is quite high.
- The biota is specialized to live with flow conditions.

Abiotic Factors

Flow

Water flow is the key factor in lotic systems influencing their ecology. The strength of water flow can vary between systems, ranging from torrential rapids to slow backwaters that almost seem like lentic systems. The speed of the water flow can also vary within a system and is subject to chaotic turbulence.

This turbulence results in divergences of flow from the mean downslope flow vector as typified by eddy currents. The mean flow rate vector is based on variability of friction with the bottom or sides of the channel, sinuosity, obstructions, and the incline gradient. In addition, the amount of water input into the system from direct precipitation, snowmelt, and/or groundwater can affect flow rate. Flowing waters can alter the shape of the streambed through erosion and deposition, creating a variety of habitats, including riffles, glides, and pools.

Light

Light is important to lotic systems, because it provides the energy necessary to drive primary production via photosynthesis, and can also provide refuge for prey species in shadows it casts. The amount of light that a system receives can be related to a combination of internal and external stream variables. The area surrounding a small stream, for example, might be shaded by surrounding forests or by valley walls. Larger river systems tend to be wide so the influence of external variables is minimized, and the sun reaches the surface.

These rivers also tend to be more turbulent, however, and particles in the water increasingly attenuate light as depth increases. Seasonal and diurnal factors might also play a role in light availability because the angle of incidence, the angle at which light strikes water can lead to light lost from reflection. Known as Beer's Law, the shallower the angle, the more light is reflected and the amount of solar radiation received declines logarithmically with depth. Additional influences on light availability include cloud cover, altitude, and geographic position (Brown 1987).

Temperature

Most lotic species are poikilotherms whose internal temperature varies with their environment, thus temperature is a key abiotic factor for them. Water can be heated or cooled through radiation at the surface and conduction to or from the air and surrounding substrate. Shallow streams are typically well mixed and maintain a relatively uniform temperature within an area. In deeper, slower moving water systems, however, a strong difference between the bottom and surface temperatures may develop. Spring fed systems have little variation as springs are typically from groundwater sources, which are often very close to ambient temperature. Many systems show strong diurnal fluctuations and seasonal variations are most extreme in arctic, desert and temperate systems. The amount of shading, climate and elevation can also influence the temperature of lotic systems.

Chemistry

Water chemistry between systems varies tremendously. The chemistry is foremost determined by inputs from the geology of its watershed, or catchment area, but can also be influenced by precipitation and the addition of pollutants from human sources. Large differences in chemistry do not usually exist within small lotic systems due to a high rate of mixing. In larger river systems, however, the concentrations of most nutrients, dissolved salts, and pH decrease as distance increases from the river's source.

Oxygen is likely the most important chemical constituent of lotic systems, as all aerobic organisms require it for survival. It enters the water mostly via diffusion at the water-air interface. Oxygen's solubility in water decreases as water pH and temperature increases.

Fast, turbulent streams expose more of the water's surface area to the air and tend to have low temperatures and thus more oxygen than slow, backwaters. Oxygen is a byproduct of photosynthesis, so systems with a high abundance of aquatic algae and plants may also have high concentrations of oxygen during the day. These levels can decrease significantly during the night when primary producers switch to respiration. Oxygen can be limiting if circulation between the surface and deeper layers is poor, if the activity of lotic animals is very high, or if there is a large amount of organic decay occurring.

Substrate

The inorganic substrate of lotic systems is composed of the geologic material present in the catchment that is eroded, transported, sorted, and deposited by the current. Inorganic substrates are classified by size on the Wentworth scale, which ranges from boulders, to pebbles, to gravel, to sand, and to silt. Typically, particle size decreases downstream with larger boulders and stones in more mountainous areas and sandy bottoms in lowland rivers. This is because the higher gradients of mountain streams facilitate a faster flow, moving smaller substrate materials further downstream for deposition. Substrate can also be organic and may include fine particles, autumn shed leaves, submerged wood, moss, and more evolved plants. Substrate deposition is not necessarily a permanent event, as it can be subject to large modifications during flooding events.

Biotic Factors

Bacteria

Bacteria are present in large numbers in lotic waters. Free-living forms are associated with decomposing organic material, biofilm on the surfaces of rocks and vegetation, in between particles that compose the substrate, and suspended in the water column. Other forms are also associated with the guts of lotic organisms as parasites or in commensal relationships. Bacteria play a large role in energy recycling, which will be discussed in the Trophic Relationships section.

Primary Producers

Algae, consisting of phytoplankton and periphyton, are the most significant sources of primary production in most streams and rivers. Phytoplankton float freely in the water column and thus are unable to maintain populations in fast flowing streams. They can, however, develop sizable populations in slow moving rivers and backwaters. Periphyton are typically filamentous and tufted algae that can attach themselves to objects to avoid being washed away by fast current. In places where flow rates are negligible or absent, periphyton may form a gelatinous, unanchored floating mat.

Plants exhibit limited adaptations to fast flow and are most successful in reduced currents. More primitive plants, such as mosses and liverworts attach themselves to solid objects. This typically occurs in colder headwaters where the mostly rocky substrate offers attachment sites. Some plants are free floating at the water's surface in dense mats like duckweed or water hyacinth. Others are rooted and may be classified as submerged or emergent. Rooted plants usually occur in areas of slackened current where fine-grained soils are found. These rooted plants are flexible, with elongated leaves that offer minimal resistance to current.

Living in flowing water can be beneficial to plants and algae because the current is usually well aerated and it provides a continuous supply of nutrients. These organisms are limited by flow, light, water chemistry, substrate, and grazing pressure. Algae and plants are important to lotic systems as sources of energy, for forming microhabitats that shelter other fauna from predators and the current, and as a food resource.

Insects and Other Invertebrates

Up to 90% of invertebrates in some lotic systems are insects. These species exhibit tremendous diversity and can be found occupying almost every available habitat, including the surfaces of stones, deep below the substratum, adrift in the current, and in the surface film. Insects have developed several strategies for living in the diverse flows of lotic systems. Some avoid high current areas, inhabiting the substratum or the sheltered side of rocks. Additional invertebrate taxa common to flowing waters include mollusks such as snails, limpets, clams, mussels, as well as crustaceans like crayfish and crabs. Like most of the primary consumers, lotic invertebrates often rely heavily on the current to bring them food and oxygen (Brown 1987). Invertebrates, especially insects, are important as both consumers and prey items in lotic systems.

Fish and Other Vertebrates

Fishes are probably the best-known inhabitants of lotic systems. The ability of a fish species to live in flowing waters depends upon the speed at which it can swim and the duration that its speed can be maintained. This ability can vary greatly between species and is tied to the habitat in which it can survive. Continuous swimming expends a tremendous amount of energy and, therefore, fishes spend only short periods in full current. Instead, individuals remain close to the bottom or the banks, behind obstacles, and sheltered from the current, swimming in the current only to feed or change locations. Some species have adapted to living only on the system bottom, never venturing into the open water flow. These fishes are dorso-ventrally flattened to reduce flow resistance and often have eyes on top of their heads to observe what is happening above them. Some also have sensory barrels positioned under the head to assist in the testing of substratum.

Lotic systems typically connect to each other, forming a path to the ocean (spring '! stream '! river '! ocean), and many fishes have life cycles that require stages in both fresh and salt water. Salmon, for example, are anadromous species that are born and develop in freshwater and then move to the ocean as adults. Eels are catadromous, and are born and develop in the ocean and then move into freshwater as adults.

Other vertebrate taxa that inhabit lotic systems include amphibians, such as salamanders, reptiles (e.g. snakes, turtles, crocodiles and alligators) various bird species, and mammals (e.g., otters, beavers, hippos, and river dolphins). With the exception of a few species, these vertebrates are not tied to water as fishes are, and spend part of their time in terrestrial habitats. Many fish species are important as consumers and as prey species to the larger vertebrates mentioned above.

Trophic Relationships

Energy Inputs

Energy sources can be autochthonous or allochthonous.

- Autochthonous energy sources are those derived from within the lotic system. During photosynthesis, for example, primary producers form organic carbon compounds out of carbon dioxide and inorganic matter. The energy they produce is important for the community because it may be transferred to higher trophic levels via consumption. Additionally, high rates of primary production can introduce dissolved organic matter (DOM) to the waters. Another form of autochthonous energy comes from the decomposition of dead organisms and feces that originate within the lotic system. In this case, bacteria decompose the detritus or coarse particulate organic material (CPOM; >1 mm pieces) into fine particulate organic matter (FPOM; <1 mm pieces) and then further into inorganic compounds that are required for photosynthesis. This process is discussed in more detail below.

- Allochthonous energy sources are those derived from outside the lotic system, that is, from the terrestrial environment. Leaves, twigs, fruits, etc. are typical forms of terrestrial CPOM that have entered the water by direct litterfall or lateral leaf blow. In addition, terrestrial animal-derived materials, such as feces or carcasses that have been added to the system are examples of allochthonous CPOM. The CPOM undergoes a specific process of degradation. Allan gives the example of a leaf fallen into a stream. First, the soluble chemicals are dissolved and leached from the leaf upon its saturation with water. This adds to the DOM load in the system. Next, microbes such as bacteria and fungi colonize the leaf, softening it as the mycelium of the fungus grows into it. The composition of the microbial community is influenced by the species of tree from which the leaves are shed (Rubbo and Kiesecker 2004). This combination of bacteria, fungi, and leaf are a food source for shredding invertebrates, which leave only FPOM after consumption. These fine particles may be colonized by microbes again or serve as a food source for animals that consume FPOM. Organic matter can also enter the lotic system already in the FPOM stage by wind, surface runoff, bank erosion, or groundwater. Similarly, DOM can be introduced through canopy drip from rain or from surface flows.

Invertebrates

Invertebrates can be organized into many feeding guilds in lotic systems. Some species are shredders, which use large and powerful mouth parts to feed on non-woody CPOM and their associated microorganisms. Others are suspension feeders, which use their setae, filtering aparati, nets, or even secretions to collect FPOM and microbes from the water. These species may be passive collectors, utilizing the natural flow of the system, or they may generate their own current to draw water, and also, FPOM in Allan. Members of the gatherer-collector guild actively search for FPOM under rocks and in other places where the stream flow has slackened enough to allow deposition. Grazing invertebrates utilize scraping, rasping, and browsing adaptations to feed on periphyton and detritus. Finally, several families are predatory, capturing and consuming animal prey. Both the number of species and the abundance of individuals within each guild is largely dependent upon food availability. Thus, these values may vary across both seasons and systems.

Fish

Fish can also be placed into feeding guilds. Planktivores pick plankton out of the water column. Herbivore-detritivores are bottom-feeding species that ingest both periphyton and detritus indiscriminately. Surface and water column feeders capture surface prey (mainly terrestrial and emerging insects) and drift (benthic invertebrates floating downstream). Benthic invertebrate feeders prey primarily on immature insects, but will also consume other benthic invertebrates. Top predators consume fishes and/or large invertebrates. Omnivores ingest a wide range of prey. These can be floral, faunal, and/or detrital in nature. Finally, parasites live off of host species, typically other fishes. Fish are flexible in their feeding roles, capturing different prey with regard to seasonal availability and their own developmental stage. Thus, they may occupy multiple feeding guilds in their lifetime. The number of species in each guild can vary greatly between systems, with temperate warm water streams having the most benthic invertebrate feeders, and tropical systems having large numbers of detritus feeders due to high rates of allochthonous input.

Community Patterns and Diversity

Local Species Richness

Large rivers have comparatively more species than small streams. Many relate this pattern to the greater area and volume of larger systems, as well as an increase in habitat diversity. Some systems, however, show a poor fit between system size and species richness. In these cases, a combination of factors such as historical rates of speciation and extinction, type of substrate, microhabitat availability, water chemistry, temperature, and disturbance such as flooding seem to be important.

Resource Partitioning

Although many alternate theories have been postulated for the ability of guild-mates to coexist, resource partitioning has been well documented in lotic systems as a means of reducing competition. The three main types of resource partitioning include habitat, dietary, and temporal segregation.

Habitat segregation was found to be the most common type of resource partitioning in natural systems. In lotic systems, microhabitats provide a level of physical complexity that can support a diverse array of organisms. The separation of species by substrate preferences has been well documented for invertebrates. Ward (1992) was able to divide substrate dwellers into 6 broad assemblages, including those that live in: coarse substrate, gravel, sand, mud, woody debris, and those associated with plants, showing one layer of segregation. On a smaller scale, further habitat partitioning can occur on or around a single substrate, such as a piece of gravel. Some invertebrates prefer the high flow areas on the exposed top of the gravel, while others reside in the crevices between one piece of gravel and the next, while still others live on the bottom of this gravel piece.

Dietary segregation is the second-most common type of resource partitioning. High degrees of morphological specializations or behavioural differences allow organisms to use specific resources. The size of nets built by some species of invertebrate suspension feeders, for example, can filter varying particle size of FPOM from the water (Edington et al. 1984). Similarly, members in the grazing guild can specialize in the harvesting of algae or detritus depending upon the morphology of their scraping apparatus. In addition, certain species seem to show a preference for specific algal species.

Temporal segregation is a less common form of resource partitioning, but it is nonetheless an observed phenomenon. Typically, it accounts for coexistence by relating it to differences in life history patterns and the timing of maximum growth among guild mates. Tropical fishes in Borneo, for example, have shifted to shorter life spans in response to the ecological niche reduction felt with increasing levels of species richness in their ecosystem.

Persistence and Succession

Over long time scales, there is a tendency for species composition in pristine systems to remain in a stable state. This has been found for both invertebrate and fish species. On shorter times scales, however, flow variability and unusual precipitation patterns decrease habitat stability and can all lead to declines in persistence levels. The ability to maintain this persistence over long time scales is related to the ability of lotic systems to return to the original community configuration relatively quickly after a disturbance (Townsend et al. 1987). This is one example of temporal succession, a site-specific change in a community involving changes in species composition over time. Another form of temporal succession might occur when a new habitat is opened up for colonization. In these cases, an entirely new community that is well adapted to the conditions found in this new area can establish itself.

River Continuum Concept

The River continuum concept (RCC) was an attempt to construct a single framework to describe the function of temperate lotic ecosystems from the source to the end and relate it to changes in the biotic community (Vannote et al. 1980). The physical basis for RCC is size and location along the gradient from a small stream eventually linked to a large river. Stream order is used as the physical measure of the position along the RCC.

According to the RCC, low ordered sites are small shaded streams where allochthonous inputs of CPOM are a necessary resource for consumers. As the river widens at mid-ordered sites, energy inputs should change. Ample sunlight should reach the bottom in these systems to support significant periphyton production. Additionally, the biological processing of CPOM (Coarse Particulate Organic Matter - larger than 1 mm) inputs at upstream sites is expected to result in the transport of large amounts of FPOM (Fine Particulate Organic Matter - smaller than 1 mm) to these downstream ecosystems. Plants should become more abundant at edges of the river with increasing river size, especially in lowland rivers where finer sediments have been deposited and facilitate rooting. The main channels likely have too much current and turbidity and a lack of substrate to support plants or periphyton. Phytoplankton should produce the only autochthonous inputs here, but photosynthetic rates will be limited due to turbidity and mixing. Thus, allochthonous inputs are expected to be the primary energy source for large rivers. This FPOM will come from both upstream sites via the decomposition process and through lateral inputs from floodplains.

Biota should change with this change in energy from the headwaters to the mouth of these systems. Namely, shredders should prosper in low-ordered systems and grazers in mid-ordered sites. Microbial decomposition should play the largest role in energy production for low-ordered sites and large rivers, while photosynthesis, in addition to degraded allochthonous inputs from upstream will be essential in mid-ordered systems. As mid-ordered sites will theoretically receive the largest variety of energy inputs, they might be expected to host the most biological diversity (Vannote et al. 1980).

Just how well the RCC actually reflects patterns in natural systems is uncertain and its generality can be a handicap when applied to diverse and specific situations. The most noted criticisms of the RCC are:

1. It focuses mostly on macroinvertebrates, disregarding that plankton and fish diversity is highest in high orders;
2. It relies heavily on the fact that low ordered sites have high CPOM inputs, even though many streams lack riparian habitats;
3. It is based on pristine systems, which rarely exist today; and
4. It is centred around the functioning of temperate streams.

Despite its shortcomings, the RCC remains a useful idea for describing how the patterns of ecological functions in a lotic system can vary from the source to the mouth.

Disturbances such as congestion by dams or natural events such as shore flooding are not included in the RCC model. Various researchers have since expanded the model to account for such irregularities. For example, J.V. Ward and J.A. Stanford came up with the Serial Discontinuity Concept in 1983, which addresses the impact of geomorphologic disorders such as congestion and integrated inflows.

The same authors presented the Hyporheic Corridor concept in 1993, in which the vertical (in depth) and lateral (from shore to shore) structural complexity of the river were connected. The flood pulse concept, developed by W.J. Junk in 1989, further modified by P.B. Bayley in 1990 and K. Tockner in 2000, takes into account the large amount of nutrients and organic material that makes its way into a river from the sediment of surrounding flooded land.

Human Impacts

Pollution

Pollutant sources of lotic systems are hard to control because they derive, often in small amounts, over a very wide area and enter the system at many locations along its length. Agricultural fields often deliver large quantities of sediments, nutrients, and chemicals to nearby streams and rivers. Urban and residential areas can also add to this pollution when contaminants are accumulated on impervious surfaces such as roads and parking lots that then drain into the system. Elevated nutrient concentrations, especially nitrogen and phosphorus which are key components of fertilizers, can increase periphyton growth, which can be particularly dangerous in slow moving streams. Another pollutant, acid rain, forms from sulphur dioxide and nitrous oxide emitted from factories and power stations.

These substances readily dissolve in atmospheric moisture and enter lotic systems through precipitation. This can lower the pH of these sites, affecting all trophic levels from algae to vertebrates (Brown 1987). Mean species richness and total species numbers within a system decrease with decreasing pH.

While direct pollution of lotic systems has been greatly reduced in the United States under the government's Clean Water Act, contaminants from diffuse, non-point sources, remain a large problem.

Flow Modification

Dams alter the flow, temperature, and sediment regime of lotic systems. Additionally, many rivers are dammed at multiple locations, amplifying the impact. Dams can cause enhanced clarity and reduced variability in stream flow, which in turn cause an increase in periphyton abundance. Invertebrates immediately below a dam can show reductions in species richness due to an overall reduction in habitat heterogeneity.

Also, thermal changes can affect insect development, with abnormally warm winter temperatures obscuring cues to break egg diapause and overly cool summer temperatures leaving too few acceptable days to complete growth. Finally, dams fragment river systems, isolating previously continuous populations, and preventing the migrations of anadromous and catadromous species.

Invasive Species

Invasive species have been introduced to lotic systems through both purposeful events (e.g. stocking game and food species) as well as unintentional events (e.g. hitchhikers on boats or fishing waders). These organisms can affect natives via competition for prey or habitat, predation, habitat alteration, hybridization, or the introduction of harmful diseases and parasites.

Once established, these species can be difficult to control or eradicate, particularly because of the connectivity of lotic systems. Invasive species can be especially harmful in areas that have endangered biota, such as mussels in the Southeast United States, or those that have localized endemic species, like lotic systems west of the Rocky Mountains, where many species evolved in isolation.

The Rivers of South Assam Mizoram and Tripura in North-East India

Northeast India is the eastern-most region of India connected to East India via a narrow corridor squeezed between Nepal and Bangladesh. It comprises the contiguous Seven Sister States (Arunachal Pradesh, Assam, Manipur, Meghalaya, Mizoram, Nagaland, and Tripura), plus the Himalayan state of Sikkim. These states are grouped under the MDONER ministry of the Government of India. Except for the Goalpara region of Assam, the rest were late entrants to the British India, the Brahmaputra valley area of Assam became a part of British India in 1824, and the hill regions were incorporated even later. Sikkim was annexed to the Indian union through a referendum in 1975 and was recognised as part of Northeast India in the 1990s.

The rivers of the North East being hidden in the Himalayas are insufficiently studied and the value of the massive river systems emerging from the Tibetan plateau, the water tower of Asia is yet to be fully understood. The region with just five per cent of the country's area has thirty per cent of the national water resource potential and its per capita water availability is 18,400 cubic meters as compared to the national figure of 2,208 cubic meters.

However, in spite of the large water potential, the region is caught in a paradox marked by the presence of some of the most water-starved pockets and bleak irrigation and agricultural development. Partial and uncertain accessibility of water resource database has resulted in lack of adequate scientific planning. Besides, water resources related institutional arrangements are bureaucratically complex and or incomplete.

The north-eastern region of India is considered as one of the hotspots of freshwater fish biodiversity in the world. This rich diversity of the region could be assigned to certain reasons, notably, the geo-morphology and the tectonics of this zone. The region is dominated by two drainage systems namely the Brahmaputra in the north and the Barak in the south. Important rivers of the Brahmaputra drainage are Subansiri, Kameng, Bhareli, Dhansiri, Manas, Umiam, Kopili, Myntang, Jingiram and Simsang while Irang, Makru, Tuivai, Jiri, Chiri, Madhura, Jatinga, Katakhal, Dhaleshwari, Singla, Longai, Sonai, Kynshi, Umngot, Myntdu and Ganol are the prominent rivers of the Barak drainage. The other hill-stream tributaries of river Barak are Bubankhal, Monkhal, Kotaikhal, Chotirkhal, Si-vadung, Lalmatikhal, Chandikhal etc.

Out of about 930 species of fishes inhabiting the freshwaters of India, the north-eastern India is represented by 267 species belonging to 114 genera under 38 families and 10 orders.

Brahmaputra: Geo-strategic aspects

The focus of the lecture was on Brahmaputra, bearing in mind that it is the real integrator of the diverse rivers of the North East. Also known as the lifeline of the region it is a major international river with a drainage area of 580,000 square kilometers, 50 per cent of which lies in China where it originates, 34 per cent in India where it surges through Arunachal Pradesh and Assam, and 8 per cent in Bangladesh which it finally touches before meeting the Bay of Bengal. Another 8 per cent of its catchment area lies in Bhutan. The first 1,625 kilometers of the Brahmaputra lies in China, the next 918 kilometers in India, and the remaining 363 kilometers in Bangladesh.

The river's origin in China makes it strategically important, more so now because it has plans to build a mega structure upstream of the Great Bend of the Yarlung Tsangpo (Brahmaputra in China) to divert water to the Gobi Desert and northern China. Because the river gains in volume only in India, to a certain extent India's fears regarding reduced inflows can be put to rest at least for monsoon flows, though this may have a negative effect on the non-monsoon flows. Dr Mahanta mentioned that regional platforms encompassing basin countries are essential for discussing these critical transnational river issues.

Brahmaputra basin in India is shared by Arunachal Pradesh (42 per cent), Assam (36 per cent), Nagaland (6 per cent), Meghalaya (6 per cent), Sikkim (4 per cent), and West Bengal (6 per cent). Water conflicts have surfaced in this ecologically and seismologically sensitive region and it has witnessed several anti-dam movements in the recent past.

The Brahmaputra faces the world's highest loss of channel gradient just before entering India (2500 m over a distance of only 200 kms). This is in the fragile young Himalayas comprised of unconsolidated sedimentary rocks marked by heavy erosion. As a result, this is the highest sediment load carrying river in the world. Lot of soft rocks and debris are carried by the fast flowing river which gets deposited on the way before reaching Bangladesh where it attains a sluggish pace.

Golden Mahseer

Mahseer is a Local words means 'The Big Mouth'; it is a fish closely related to European Barbel & Carp. This creature is known to be the toughest among the fresh water sport fish. It has an elongated body with a pointed snout. The inter-orbital space is flat while the jaws are of equal sizes. It has barbels in two pairs with the rostral barbel shorter than the maxillary. Its is a highly finicky & moody fish. The body colour is golden, with grey on the dorsal side and reddish - yellow fins. It is a cold water fish. Mahseer's habitat must have highly oxygenated water with a rocky river bed making the fast flowing rivers of northeast India ideal.

Fishing Methods in the Rivers of Northeast India

The seven Northeastern states of India, Arunachal Pradesh, Assam, Manipur, Meghalaya, Mizoram, Nagaland and Tripura are blessed with diversified fish fauna, which assumes special interest due to the fact that the region is drained by two important drainages, viz., Brahmaputra- Barak and Chindwin Irrawaddy. This region has been ranked 6th among the top 25 biodiversity spots in the world. It has varied physiographic features ranging from river valleys to snow capped mountains making it a favourable habitat for a variety of ichthyofauna. Despite such a rich resource potential, the region is lagging behind in its exploitation and management. In the high altitude stretches, utilization of rivers and streams is low due to biotic, abiotic and socio-economic constraints. Still a substantial part of the resources in this region remain unutilized and unavailable. A substantial part of the resource in the northeast comprises capture fisheries, the management of which requires higher skill. Little research attention has been given for development of fisheries in contrast to the other water bodies.

Riverine fisheries resources of Northeastern states comprise 19,150 km of streams and rivers. Except the Brahmaputra and Barak valleys in Assam and Imphal valley in Manipur, the Northeast mainly comprises hills and mountains, which form 65% of its territory. The mighty Brahmaputra and Barak along with their tributaries form more than half of rivers in this region.

Rivers of this region consist of both torrential and plain rivers. In Arunachal Pradesh, the entire 2,000 km of rivers are of true rheophillic type. Assam alone has 5,050 km of rivers including all the tributaries of Brahmaputra and Barak. Manipur has two major rivers viz, the Barak and the Manipur with a combined length of 2,000 km. There are 21 rivers in the hilly state of Mizoram, which along with their tributaries run into 1700 km. Meghalaya's 5,600 km of rivers mainly comprise the tributaries of Brahmaputra. Barak and Tizu are the main rivers of Nagaland, with a combined length of 1,600 km. Rivers of Tripura have a total length of 1,200 km. Ichthyofaunistic resources of the region exhibit a combination of both torrential and plain water forms as well as cold and warm water species and occupy diverse ecological regimes in their distributional range. The diversified aquatic ecosystem of the region harbours altogether, 230 fish species belonging to 103 genera and 36 families.

Endemic species recorded from these states are 12 each from Assam and Arunachal Pradesh, 7 from Manipur, 6 from Meghalaya and one each from Mizoram and Tripura.

Hydrotopography

Hydrography is the branch of applied sciences which deals with the measurement and description of the physical features of oceans, seas, coastal areas, lakes and rivers, as well as with the prediction of their change over time, for the primary purpose of safety of navigation and in support of all other marine activities, including economic development, security and defence, scientific research, and environmental protection.

The origins of hydrography lay in the making of charts to aid navigation, by individual mariners as they navigated into new waters. These were usually the private property, even closely held secrets, of individuals who used them for commercial or military advantage. As transoceanic trade and exploration increased, hydrographic surveys started to be carried out as an exercise in their own right, and the commissioning of surveys was increasingly done by governments and special hydrographic offices. National organizations, particularly navies, realized that the collection, systematization and distribution of this knowledge gave it great organizational and military advantages. Thus were born dedicated national hydrographic organizations for the collection, organization, publication and distribution of hydrography incorporated into charts and sailing directions.

Large-scale hydrography is usually undertaken by national or international organizations which sponsor data collection through precise surveys and publish charts and descriptive material for navigational purposes. The science of oceanography is, in part, an outgrowth of classical hydrography. In many respects the data are interchangeable, but marine hydrographic data will be particularly directed toward marine navigation and safety of that navigation. Marine resource exploration and exploitation is a significant application of hydrography, principally focused on the search for hydrocarbons.

Hydrographical measurements include the tidal, current and wave information of physical oceanography. They include bottom measurements, with particular emphasis on those marine geographical features that pose a hazard to navigation such as rocks, shoals, reefs and other features that obstruct ship passage. Bottom measurements also include collection of the nature of the bottom as it pertains to effective anchoring. Unlike oceanography, hydrography will include shore features, natural and manmade, that aid in navigation. Therefore, a hydrographic survey may include the accurate positions and representations of hills, mountains and even lights and towers that will aid in fixing a ship's position, as well as the physical aspects of the sea and seabed.

Hydrography, mostly for reasons of safety, adopted a number of conventions that have affected its portrayal of the data on nautical charts. For example, hydrographic charts are designed to portray what is safe for navigation, and therefore will usually tend to maintain least depths and occasionally de-emphasize the actual submarine topography that would be portrayed on bathymetric charts. The former are the mariner's tools to avoid accident. The latter are best representations of the actual seabed, as in a topographic map, for scientific and other purposes. Trends in hydrographic practice since c. 2003–2005 have led to a narrowing of this difference, with many more hydrographic offices maintaining "best observed" databases, and then making navigationally "safe" products as required. This has been coupled with a preference for multi-use surveys, so that the same data collected for nautical charting purposes can also be used for bathymetric portrayal.

Even though, in places, hydrographic survey data may be collected in sufficient detail to portray bottom topography in some areas, hydrographic charts only show depth information relevant for safe navigation and should not be considered as a product that accurately portrays the actual shape of the bottom. The soundings selected from the raw source depth data for placement on the nautical chart are selected for safe navigation and are biased to show predominately the shallowest depths that relate to safe navigation. For instance, if there is a deep area that can not be reached because it is surrounded by shallow water, the deep area may not be shown. The colour filled areas that show different ranges of shallow water are not the equivalent of contours on a topographic map since they are often drawn seaward of the actual shallowest depth portrayed. A bathymetric chart does show marine topology accurately. Details covering the above limitations can be found in Part 1 of Bowditch's American Practical Navigator. Another concept that affects safe navigation is the sparsity of detailed depth data from high resolution sonar systems. In more remote areas, the only available depth information has been collected with lead lines. This collection method drops a weighted line to the bottom at intervals and records the depth, often from a rowboat or sail boat. There is no data between soundings or between sounding lines to guarantee that there is not a hazard such as a wreck or a coral head waiting there to ruin a sailor's day. Often, the navigation of the collecting boat does not match today's GPS navigational accuracies. The hydrographic chart will use the best data available and will caveat its nature in a caution note or in the legend of the chart.

A hydrographic survey is quite different from a bathymetric survey in some important respects, particularly in a bias toward least depths due to the safety requirements of the former and geomorphologic descriptive requirements of the latter. Historically, this could include echosoundings being conducted under settings biased toward least depths, but in modern practice hydrographic surveys typically attempt to best measure the depths observed, with the adjustments for navigational safety being applied after the fact.

Hydrography of streams will include information on the stream bed, flows, water quality and surrounding land. Basin or interior hydrography pays special attention to rivers and potable water although if collected data is not for ship navigational uses, and is intended for scientific usage, it is more commonly called *hydrology*.

Hydrography of rivers and streams is also an integral part of water management. Most reservoirs in the United States use dedicated stream gauging and rating tables to determine inflows into the reservoir and outflows to irrigation districts, water municipalities and other users of captured water. River/stream hydrographers use handheld and bank mounted devices, to capture a sectional flow rate of moving water through a section.

When viewing a topographic map, one of the features immediately visible is a simple definition of the character of the land surface. Forested areas are shown as green, water areas as blue, cities as black, and so on. The way the landscape responds to rainfall depends a great deal on land cover, and to some degree also on other factors, such as soil type. These factors may collectively be represented as Hydro Response Units, a classification of the land surface into a set of polygons each considered homogeneous in its vertical water balance properties, that is, the properties that define how precipitation will be transformed into runoff, evaporation and groundwater recharge at a point on the land surface.

Data layers describing map hydrography, point data inventories and hydro response units make up the Hydrography component of the ArcGIS Hydro data model.

Features are the parent class for Hydro Feature and from there Hydro Feature is the parent class for Hydrography. A feature is any row in a table that has a unique identifier and a shape. Shapes can be 0-dimensional (points), 1-dimensional (lines) or 2-dimensional (areas). In this case, we have a feature class for each dimension, Hydro Points, Hydro Lines, and Hydro Areas. The Hydro Point class has seven child classes, Monitoring Point, Dam, Bridge, Structure, Water Withdrawal, Water Discharge, and User Point. The User Point class is meant to hold point features that do not fit into the other classes. Hydro Area contains one child class, Water-body.

Morphology, Morphometry and Hydrodynamics

Hydromorphology is the study of the physical characteristics of bodies of water on the Earth's surface, including river basins, channels, streams, and lakes. Water quality, levels of pollution, and biological components needed for ecological system maintenance are a few areas assessed when classifying water systems. Hydromorphology studies the dynamics of groundwater flow into channels, lakes, and streams. It measures flow patterns and geometry as well as routing flows to avoid flooding or drought.

A body of water or waterbody (often spelled water body) is any significant accumulation of water, generally on a planet's surface. The term *body of water* most often refers to large accumulations of water, such as oceans, seas, and lakes, but it includes smaller pools of water such as ponds, wetlands, or more rarely, puddles. A body of water does not have to be still or contained; Rivers, streams, canals, and other geographical features where water moves from one place to another are also considered bodies of water.

Most are naturally occurring geographical features, but some are man-made (artificial). There are types that can be either. For example, most reservoirs are created by engineering dams, but some natural lakes are used as reservoirs. Similarly, most harbours are naturally occurring bays, but some harbours have been created through construction.

Bodies of water that are navigable are known as waterways. Some bodies of water collect and move water, such as rivers and streams, and others primarily hold water, such as lakes and oceans.

The term *body of water* can also refer to a reservoir of water held by a plant, technically known as a phytotelma.

Morphometry

Morphometrics or morphometry refers to the quantitative analysis of *form*, a concept that encompasses size and shape.

Morphometric analyses are commonly performed on organisms, and are useful in analysing their fossil record, the impact of mutations on shape, developmental changes in form, covariances between ecological factors and shape, as well for estimating quantitative-genetic parameters of shape.

Morphometrics can be used to quantify a trait of evolutionary significance, and by detecting changes in the shape, deduce something of their ontogeny, function or evolutionary relationships. A major objective of morphometrics is to statistically test hypotheses about the factors that affect shape.

Three general approaches to form are usually distinguished: traditional morphometrics, landmark-based morphometrics and outline-based morphometrics.

Species identification and population discrimination are important in the conservation of biodiversity, natural resources, and fisheries management. It is also often necessary to identify individual specimens correctly to population when investigating biological traits such as growth, mortality, fecundity, trophic relationships, parasite relationships, and historical and paleontological events.

Preliminary visual assessment was used to identify potential landmarks on the scales. The landmarks chosen are located on key features of the scale that are common to all scales of the species or variants examined. This ensures that in subsequent interpretation of results, variations in particular landmarks can be related back to shared features of form (contrast with Fourier analysis where no such relationship exists except for the start point for the series).

For mullet scales, the following landmarks were considered appropriate: landmarks 1 and 3 are the ventro- and dorso-lateral tips of the anterior portion of the scale, respectively; landmark 2 is in the centre of the anterior edge of the scale, landmarks 4 and 6 are at the boundary between the anterior area with circuli and the posterior area covered by cteni (spine-like ornamentations), respectively; landmark 5 is the focus of the scale; and landmark 7 is positioned at the tip of the posterior portion of the scale.

Hydrodynamics

To understand the mechanics of fish swimming, we need to know the forces exerted by the fluid and how these forces affect the motion of the fish. To this end, we developed a 3-D computational approach that integrates hydrodynamics and body dynamics.

Fish swim by exerting force against the surrounding water. There are exceptions, but this is normally achieved by the fish contracting muscles on either side of its body in order to generate waves of flexion that travel the length of the body from nose to tail, generally getting larger as they go along. The vector forces exerted on the water by such motion cancel out laterally, but generate a net force backwards which in turn pushes the fish forward through the water.

Most fishes generate thrust using lateral movements of their body and caudal fin. But there are also a huge number of species that move mainly using their median and paired fins. The latter group profits from the gained maneuverability that is needed when living in coral reefs for example. But they can't swim as fast as fish using their bodies and caudal fins.

Bone and muscle tissues of fish are denser than water. To maintain depth some fish increase buoyancy by means of a gas bladder or by storing oils or lipids. Fish without these features use dynamic lift instead. It is done using their pectoral fins in a manner similar to the use of wings by airplanes and birds. As these fish swim, their pectoral fins are positioned to create lift which allows the fish to maintain a certain depth.

Sharks are a notable example of fish that depend on dynamic lift; notice their well-developed pectoral fins.

The two major drawbacks of this method are that these fish must stay moving to stay afloat and that they are incapable of swimming backwards or hovering.

Hydrodynamic Principles

Similarly to the aerodynamics of flight, powered swimming requires animals to overcome drag by producing thrust. Unlike flying, however, swimming animals do not necessarily need to actively exert high vertical forces because the effect of buoyancy can counter the downward pull of gravity, allowing these animals to float without much effort. While there is great diversity in fish locomotion, swimming behaviour can be classified into two distinct "modes" based on the body structures involved in thrust production, Median-Paired Fin (MPF) and Body-Caudal Fin (BCF). Within each of these classifications, there are a numerous specifications along a spectrum of behaviours from purely undulatory to entirely oscillatory based.

In undulatory swimming modes thrust is produced by wave-like movements of the propulsive structure (usually a fin or the whole body). Oscillatory modes, on the other hand, are characterized by thrust production from swiveling of the propulsive structure on an attachment point without any wave-like motion.

Physico-chemical Characteristics of Soil

Soil moisture limits forage production potential the most in semiarid regions. Estimated water use efficiency for irrigated and dry-land crop production systems is 50 percent, and available soil water has a large impact on management decisions producers make throughout the year. Soil moisture available for plant growth makes up approximately 0.01 percent of the world's stored water.

By understanding a little about the soil's physical properties and its relationship to soil moisture, you can make better soil-management decisions. Soil texture and structure greatly influence water infiltration, permeability, and water-holding capacity.

Soil texture refers to the composition of the soil in terms of the proportion of small, medium, and large particles (clay, silt, and sand, respectively) in a specific soil mass. For example, a coarse soil is a sand or loamy sand, a medium soil is a loam, silt loam, or silt, and a fine soil is a sandy clay, silty clay, or clay.

Soil structure refers to the arrangement of soil particles (sand, silt, and clay) into stable units called aggregates, which give soil its structure. Aggregates can be loose and friable, or they can form distinct, uniform patterns. For example, granular structure is loose and friable, blocky structure is six-sided and can have angled or rounded sides, and platelike structure is layered and may indicate compaction problems.

Soil porosity refers to the space between soil particles, which consists of various amounts of water and air. Porosity depends on both soil texture and structure.

For example, a fine soil has smaller but more numerous pores than a coarse soil. A coarse soil has bigger particles than a fine soil, but it has less porosity, or overall pore space. Water can be held tighter in small pores than in large ones, so fine soils can hold more water than coarse soils.

Water infiltration is the movement of water from the soil surface into the soil profile. Soil texture, soil structure, and slope have the largest impact on infiltration rate. Water moves by gravity into the open pore spaces in the soil, and the size of the soil particles and their spacing determines how much water can flow in. Wide pore spacing at the soil surface increases the rate of water infiltration, so coarse soils have a higher infiltration rate than fine soils.

Permeability refers to the movement of air and water through the soil, which is important because it affects the supply of root-zone air, moisture, and nutrients available for plant uptake. A soil's permeability is determined by the relative rate of moisture and air movement through the most restrictive layer within the upper 40 inches of the effective root zone. Water and air rapidly permeate coarse soils with granular subsoils, which tend to be loose when moist and don't restrict water or air movement. Slow permeability is characteristic of a moderately fine subsoil with angular to subangular blocky structure. It is firm when moist and hard when dry.

Water-holding capacity is controlled primarily by soil texture and organic matter. Soils with smaller particles (silt and clay) have a larger surface area than those with larger sand particles, and a large surface area allows a soil to hold more water. In other words, a soil with a high percentage of silt and clay particles, which describes fine soil, has a higher water-holding capacity. Organic matter percentage also influences water-holding capacity. As the percentage increases, the water-holding capacity increases because of the affinity organic matter has for water.

Water availability is illustrated in the figure by water levels in three different soil types. Excess or gravitational water drains quickly from the soil after a heavy rain because of gravitational forces (saturation point to field capacity). Plants may use small amounts of this water before it moves out of the root zone. Available water is retained in the soil after the excess has drained (field capacity to wilting point). This water is the most important for crop or forage production. Plants can use approximately 50 percent of it without exhibiting stress, but if less than 50 percent is available, drought stress can result. Unavailable water is soil moisture that is held so tightly by the soil that it cannot be extracted by the plant. Water remains in the soil even below plants' wilting point.

Soil texture greatly influences water availability. The sandy soil can quickly be recharged with soil moisture but is unable to hold as much water as the soils with heavier textures. As texture becomes heavier, the wilting point increases because fine soils with narrow pore spacing hold water more tightly than soils with wide pore spacing.

Soil is a valuable resource that supports plant life, and water is an essential component of this system. Management decisions concerning types of crops to plant, plant populations, irrigation scheduling, and the amount of nitrogen fertilizer to apply depend on the amount of moisture that is available to the crop throughout the growing season. By understanding some physical characteristics of the soil, you can better define the strengths and weaknesses of different soil types.

Plankton Communities

Plankton (singular plankter) are a diverse group of organisms that live in the water column and cannot swim against a current. They provide a crucial source of food to many large aquatic organisms, such as fish and whales.

These organisms include drifting animals, protists, archaea, algae, or bacteria that inhabit the pelagic zone of oceans, seas, or bodies of fresh water; that is, plankton are defined by their ecological niche rather than phylogenetic or taxonomic classification.

Though many planktic (or planktonic) species are microscopic in size, *plankton* includes organisms covering a wide range of sizes, including large organisms such as jellyfish.

The name plankton is derived from the Greek adjective *planktos*, meaning errant, and by extension "wanderer" or *drifter*. It was coined by Victor Hensen. Plankton typically flow with ocean currents. While some forms are capable of independent movement and can swim hundreds of meters vertically in a single day (a behaviour called diel vertical migration), their horizontal position is primarily determined by the surrounding currents. This is in contrast to nekton organisms that can swim against the ambient flow and control their position (e.g. squid, fish, and marine mammals).

Within the plankton, holoplankton spend their entire life cycle as plankton (e.g. most algae, copepods, salps, and some jellyfish). By contrast, meroplankton are only planktic for part of their lives (usually the larval stage), and then graduate to either a nektic or benthic (sea floor) existence. Examples of meroplankton include the larvae of sea urchins, starfish, crustaceans, marine worms, and most fish.

Plankton abundance and distribution are strongly dependent on factors such as ambient nutrient concentrations, the physical state of the water column, and the abundance of other plankton.

The study of plankton is termed planktology and a planktonic individual is referred to as a plankter. The adjective planktonic is widely used in both the scientific and popular literature, and is a generally accepted term.

However, from the standpoint of formal grammar the less commonly used planktic is more strictly the correct adjective. When deriving English words from their Greek or Latin roots the gender specific ending (in this case "-on," which indicates the word is neuter) is normally dropped, using only the root of the word in the derivation.

Trophic Groups

Plankton are primarily divided into broad functional (or trophic level) groups:

- Phytoplankton (from Greek *phyton*, or plant), autotrophic, prokaryotic or eukaryotic algae that live near the water surface where there is sufficient light to support photosynthesis. Among the more important groups are the diatoms, cyanobacteria, dinoflagellates and coccolithophores.
- Zooplankton (from Greek *zoon*, or animal), small protozoans or metazoans (e.g. crustaceans and other animals) that feed on other plankton and telonemia. Some of the eggs and larvae of larger animals, such as fish, crustaceans, and annelids, are included here.
- Bacterioplankton, bacteria and archaea, which play an important role in remineralising organic material down the water column (note that the prokaryotic phytoplankton are also bacterioplankton).

This scheme divides the plankton community into broad producer, consumer and recycler groups. However, determining the trophic level of some plankton is not straightforward.

For example, although most dinoflagellates are either photosynthetic producers or heterotrophic consumers, many species are mixotrophic depending upon circumstances.

Size Groups

However, some of these terms may be used with very different boundaries, especially on the larger end. The existence and importance of nano- and even smaller plankton was only discovered during the 1980s, but they are thought to make up the largest proportion of all plankton in number and diversity.

The microplankton and smaller groups are microorganisms and operate at low Reynolds numbers, where the viscosity of water is much more important than its mass or inertia.

Group	Size range (ESD)	Examples
Megaplankton	> 20 cm	metazoans; *e.g.* jellyfish; ctenophores; salps and pyrosomes (pelagic Tunicata); Cephalopoda; Amphipoda
Macroplankton	2? 20 cm	metazoans; *e.g.* Pteropods; Chaetognaths; Euphausiacea (krill); Medusae; ctenophores; salps, doliolids and pyrosomes (pelagic Tunicata); Cephalopoda; Janthinidae (one family gastropods); Amphipoda
Mesoplankton	0.2? 20 mm	metazoans; *e.g.* copepods; Medusae; Cladocera; Ostracoda; Chaetognaths; Pteropods; Tunicata; Heteropoda
Microplankton	20? 200 μm	large eukaryotic protists; most phytoplankton; Protozoa Foraminifera; tintinnids; other ciliates; Rotifera; juvenile metazoans - Crustacea (copepod nauplii)
Nanoplankton	2? 20 μm	small eukaryotic protists; Small Diatoms; Small Flagellates; Pyrrophyta; Chrysophyta; Chlorophyta; Xanthophyta
Picoplankton	0.2? 2 μm	small eukaryotic protists; bacteria; Chrysophyta
Femtoplankton	< 0.2 μm	marine viruses

Distribution

Plankton inhabit oceans, seas, lakes, ponds. Local abundance varies horizontally, vertically and seasonally. The primary cause of this variability is the availability of light. All plankton ecosystems are driven by the input of solar energy, confining primary production to surface waters, and to geographical regions and seasons having abundant light.

A secondary variable is nutrient availability. Although large areas of the tropical and sub-tropical oceans have abundant light, they experience relatively low primary production because they offer limited nutrients such as nitrate, phosphate and silicate. This results from large-scale ocean circulation and water column stratification. In such regions, primary production usually occurs at greater depth, although at a reduced level (because of reduced light).

Despite significant macronutrient concentrations, some ocean regions are unproductive (so-called HNLC regions). The micronutrient iron is deficient in these regions, and adding it can lead to the formation of blooms of many kinds of phytoplankton. Iron primarily reaches the ocean through the deposition of dust on the sea surface. Paradoxically, oceanic areas adjacent to unproductive, arid land thus typically have abundant phytoplankton (e.g., the eastern Atlantic Ocean, where trade winds bring dust from the Sahara Desert in north Africa). While plankton are most abundant in surface waters, they live throughout the water column. At depths where no primary production occurs, zooplankton and bacterioplankton instead consume organic material sinking from more productive surface waters above. This flux of sinking material, so-called marine snow, can be especially high following the termination of spring blooms.

Biogeochemical Significance

Aside from representing the bottom few levels of a food chain that supports commercially important fisheries, plankton ecosystems play a role in the biogeochemical cycles of many important chemical elements, including the ocean's carbon cycle. Primarily by grazing on phytoplankton, zooplankton provide carbon to the planktic foodweb, either respiring it to provide metabolic energy, or upon death as biomass or detritus. Organic material tends to be denser than seawater, and as a result it sinks into open ocean ecosystems away from the coastlines, transporting carbon along with it. This process is known as the biological pump, and it is one reason that oceans constitute the largest carbon sink on Earth.

It might be possible to increase the ocean's uptake of carbon dioxide generated through human activities by increasing plankton production through *seeding*, primarily with the micronutrient iron. However, this technique may not be practical at a large scale. Ocean oxygen depletion and resultant methane production (caused by the excess production remineralising at depth) is one potential drawback.

Biomass Variability

The growth of phytoplankton populations is dependent on light levels and nutrient availability. The chief factor limiting growth varies from region to region in the world's oceans. On a broad scale, growth of phytoplankton in the oligotrophic tropical and subtropical gyres is generally limited by nutrient supply, while light often limits phytoplankton growth in subarctic gyres. Environmental variability at multiple scales influences the nutrient and light available for phytoplankton, and as these organisms form the base of the marine food web, this variability in phytoplankton growth influences higher trophic levels. For example, at interannual scales phytoplankton levels temporarily plummet during El Niño periods, influencing populations of zooplankton, fishes, sea birds, and marine mammals.

The effects of anthropogenic warming on the global population of phytoplankton is an area of active research. Changes in the vertical stratification of the water column, the rate of temperature-dependent biological reactions, and the atmospheric supply of nutrients are expected to have important impacts on future phytoplankton productivity. Additionally, changes in the mortality of phytoplankton due to rates of zooplankton grazing may be significant.

Freshly hatched fish larvae are also plankton for a few days as long as they cannot swim against currents.

Zooplankton are the initial prey item for almost all fish larvae as they switch from their yolk sacs to external feeding. Fish rely on the density and distribution of zooplankton to match that of new larvae, which can otherwise starve. Natural factors (e.g., current variations) and man-made factors (e.g. river dams) can strongly affect zooplankton, which can in turn strongly affect larval survival, and therefore breeding success.

Freshwater Zooplankton

Lakes and streams are teeming with life but most of these organisms cannot be seen with the naked eye.

If you haul a very fine-meshed net through the water, tip your catch into a jar and hold it up to the light, you may see some of the larger species of zooplankton swimming jerkily up and down.

Zooplankton are organisms that have animal-like traits. The biggest are only five millimetres long and the smallest are just one thousandth of this size. They float, drift or weakly swim in the water. In fact, the name plankton comes from the Greek word 'planktos' which means 'wanderer' or 'drifter'.

Zooplankton encompass a wide range of both unicellular and multicellular animals. While most zooplankton are 'heterotrophs' – that is they obtain their energy from consuming organic compounds, such as algae or other zooplankton - some zooplankton, such as the dinoflagellates, may also be fully or partially photosynthetic - gaining their energy, as plants do, from sunlight.

The design and symmetry found in this aqueous world is beautiful and enthralling.

Freshwater zooplankton are found in the water in wetland areas such as lakes, tarns, streams and swamps. They are most abundant nearer the surface as they eat phytoplankton (microscopic plants) which need light to photosynthesise.

Many species move into shallower waters at night. Avoidance of UV and/or predators and metabolic advantages are some of the possible reasons for this phenomenon.

Most plankton are too small to see with the naked eye, but their beautiful shapes are revealed under the microscope.

Dominant among the larger organisms are Cladocerans which swim by rowing with their large antennae in a series of jerks. They are covered by a transparent shell (bivalve carapace) which has to be shed periodically for them to grow bigger. *Bosmina* are abundant in most lakes.

Daphnia pulex, a recently introduced species, is much larger than any of our native Cladocerans. It can reach five millimetres in length and may be seen with the naked eye.

Zooplankton are a vital component of freshwater food webs. The smallest zooplankton are eaten by the larger zooplankton which, in turn, are eaten by small fish, aquatic insects and so on.

Herbivorous zooplankton graze on phytoplankton or algae, and help maintain the natural balance of algae.

Variations in water quality such as changes in nutrient levels, conductivity, temperature or pH, can lead to changes in species composition and abundance.

Temporary drying out or freezing of water bodies or increases in water temperature, present challenging conditions for zooplankton. Often it is only the resting egg stages that survive these periods.

Exotic invading species, such as *Daphnia pulex*, may out-compete native species, which could result in irreversible changes in food webs.

Aquatic Macrophytes

A macrophyte is an aquatic plant that grows in or near water and is either emergent, submergent, or floating. In lakes and rivers macrophytes provide cover for fish and substrate for aquatic invertebrates, produce oxygen, and act as food for some fish and wildlife.

A decline in a macrophyte community may indicate water quality problems and changes in the ecological status of the water body. Such problems may be the result of excessive turbidity, herbicides, or salinization. Conversely, overly high nutrient levels may create an overabundance of macrophytes, which may in turn interfere with lake processing.

Macrophyte levels are easy to sample, do not require laboratory analysis, and are easily used for calculating simple abundance metrics.

Aquatic plants are plants that have adapted to living in aquatic environments. They are also referred to as hydrophytes or macrophytes. These plants require special adaptations for living submerged in water, or at the water's surface. The most common adaptation is aerenchyma, but floating leaves and finely dissected leaves are also common. Aquatic plants can only grow in water or in soil that is permanently saturated with water. They are therefore a common component of wetlands.

The principal factor controlling the distribution of aquatic plants is the depth and duration of flooding. However, other factors may also control their distribution, abundance, and growth form, including nutrients, disturbance from waves, grazing, and salinity.

Aquatic vascular plants have originated on multiple occasions in different plant families; they can be ferns or angiosperms (including both monocots and dicots). Seaweeds are not vascular plants; rather they are multicellular marine algae, and therefore are not typically included among aquatic plants.

A few aquatic plants are able to survive in brackish, saline, and salt water. Examples are found in genera such as Thalassia and Zostera. Although most aquatic plants can reproduce by flowering and setting seed, many also have extensive asexual reproduction by means of rhizomes, turions, and fragments in general.

One of the largest aquatic plants in the world is the Amazon water lily; one of the smallest is the minute duckweed. Many small aquatic animals use plants like duckweed for a home, or for protection from predators.

But areas with more vegetation are likely to have more predators. Some other familiar examples of aquatic plants might include floating heart, water lily, lotus and water hyacinth.

Some aquatic plants are used by humans as a food source. Examples include wild rice (*Zizania*), water caltrop (*Trapa natans*), Chinese water chestnut (*Eleocharis dulcis*), Indian lotus (*Nelumbo nucifera*), water spinach (*Ipomoea aquatica*) and watercress (*Rorippa nasturtium-aquaticum*).

There are many possible classifications of aquatic plants based upon morphology. One example has six groups as follows:

- Amphiphytes: plants that are adapted to live either submerged or on land.
- Elodeids: stem plants that complete their entire life cycle submerged, or with only their flowers above the waterline.
- Isoetids: rosette plants that complete their entire life cycle submerged.
- Helophytes: plants rooted in the bottom, but with leaves above the waterline.
- Nymphaeids: plants rooted in the bottom, but with leaves floating on the water surface.
- Pleuston: vascular plants that float freely in the water.

Eutrophication its Problems and Consequences

Eutrophication or more precisely hypertrophication, is the ecosystem response to the addition of artificial or natural substances, mainly phosphates, through detergents, fertilizers, or sewage, to an aquatic system. One example is the "bloom" or great increase of phytoplankton in a water body as a response to increased levels of nutrients. Negative environmental effects include hypoxia, the depletion of oxygen in the water, which causes a reduction in specific fish and other animals.

Mechanism of Eutrophication

Eutrophication arises from the oversupply of nutrients, which induces explosive growth of plants and algae which, when such organisms die, consume the oxygen in the body of water, thereby creating the state of hypoxia.

According to Ullmann's Encyclopedia, "the primary limiting factor for eutrophication is phosphate." The availability of phosphorus generally promotes excessive plant growth and decay, favouring simple algae and plankton over other more complicated plants, and causes a severe reduction in water quality. Phosphorus is a necessary nutrient for plants to live, and is the limiting factor for plant growth in many freshwater ecosystems. Phosphate adheres tightly to soil, so it is mainly transported by erosion. Once translocated to lakes, the extraction of phosphate into water is slow, hence the difficulty of reversing the effects of eutrophication.

The source of this excess phosphate are detergents, industrial/ domestic run-off, and fertilizers. With the phasing out of phosphate-containing detergents in the 1970s, industrial/domestic run-off and agriculture have emerged as the dominant contributors to eutrophication.

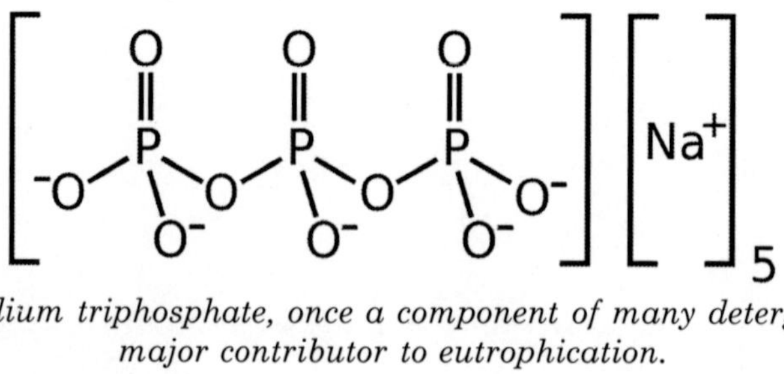

Figure: *Sodium triphosphate, once a component of many detergents, was a major contributor to eutrophication.*

Lakes and Rivers

The addition of phosphorus increases algal growth, but not all phosphates actually feed algae. Algal blooms are associated with assimilate the other necessary nutrients needed for plants and animals. When algae die they sink to the bottom where they are decomposed and the nutrients contained in organic matter are converted into inorganic form by bacteria. The decomposition process consumes oxygen, depriving fish and other organisms. Also the nutrients are concentrated at the bottom of the aquatic ecosystem and if they are not brought up closer to the surface, where there is more available light allowing for photosynthesis for aquatic plants, a serious strain is placed on algae populations.

Enhanced growth of aquatic vegetation or phytoplankton and algal blooms disrupts normal functioning of the ecosystem, causing a variety of problems such as a lack of oxygen needed for fish and shellfish to survive. The water becomes cloudy, typically coloured a shade of green, yellow, brown, or red. Eutrophication also decreases the value of rivers, lakes and aesthetic enjoyment. Health problems can occur where eutrophic conditions interfere with drinking water treatment.

Human activities can accelerate the rate at which nutrients enter ecosystems. Runoff from agriculture and development, pollution from septic systems and sewers, sewage sludge spreading, and other human-related activities increase the flow of both inorganic nutrients and organic substances into ecosystems. Elevated levels of atmospheric compounds of nitrogen can increase nitrogen availability. Phosphorus is often regarded as the main culprit in cases of eutrophication in lakes subjected to "point source" pollution from sewage pipes. The concentration of algae and the trophic state of lakes correspond well to phosphorus levels in water. Studies conducted in the Experimental Lakes Area in Ontario have shown a relationship between the addition of phosphorus and the rate of eutrophication. Humankind has increased the rate of phosphorus cycling on Earth by four times, mainly due to agricultural fertilizer production and application. Between 1950 and 1995, an estimated 600,000,000 tonnes of phosphorus were applied to Earth's surface, primarily on croplands. Policy changes to control point sources of phosphorus have resulted in rapid control of eutrophication.

Natural Eutrophication

Although eutrophication is commonly caused by human activities, it can also be a natural process, particularly in lakes. Eutrophy occurs in many lakes in temperate grasslands, for instance. Paleolimnologists now recognise that climate change, geology, and other external influences are critical in regulating the natural productivity of lakes. Some lakes also demonstrate the reverse process (meiotrophication), becoming less nutrient rich with time. The main difference between natural and anthropogenic eutrophication is that the natural process is very slow, occurring on geological time scales.

Ocean Waters

Eutrophication is a common phenomenon in coastal waters. In contrast to freshwater systems, nitrogen is more commonly the key limiting nutrient of marine waters; thus, nitrogen levels have greater importance to understanding eutrophication problems in salt water. Estuaries tend to be naturally eutrophic because land-derived nutrients are concentrated where run-off enters a confined channel. Upwelling in coastal systems also promotes increased productivity by conveying deep, nutrient-rich waters to the surface, where the nutrients can be assimilated by algae.

The World Resources Institute has identified 375 hypoxic coastal zones in the world, concentrated in coastal areas in Western Europe, the Eastern and Southern coasts of the US, and East Asia, particularly Japan. In addition to runoff from land, atmospheric fixed nitrogen can enter the open ocean. A study in 2008 found that this could account for around one third of the ocean's external (non-recycled) nitrogen supply, and up to 3% of the annual new marine biological production. It has been suggested that accumulating reactive nitrogen in the environment may prove as serious as putting carbon dioxide in the atmosphere.

Terrestrial Ecosystems

Terrestrial ecosystems are subject to similarly adverse impacts from eutrophication. Increased nitrates in soil are frequently undesirable for plants. Many terrestrial plant species are endangered as a result of soil eutrophication, such as the majority of orchid species in Europe. Meadows, forests, and bogs are characterized by low nutrient content and slowly growing species adapted to those levels, so they can be overgrown by faster growing and more competitive species. In meadows, tall grasses that can take advantage of higher nitrogen levels may change the area so that natural species may be lost. Species-rich fens can be overtaken by reed or reedgrass species. Forest undergrowth affected by run-off from a nearby fertilized field can be turned into a nettle and bramble thicket.

Chemical forms of nitrogen are most often of concern with regard to eutrophication, because plants have high nitrogen requirements so that additions of nitrogen compounds will stimulate plant growth. Nitrogen is not readily available in soil because N_2, a gaseous form of nitrogen, is very stable and unavailable directly to higher plants.

Terrestrial ecosystems rely on microbial nitrogen fixation to convert N_2 into other forms such as nitrates. However, there is a limit to how much nitrogen can be utilized. Ecosystems receiving more nitrogen than the plants require are called nitrogen-saturated. Saturated terrestrial ecosystems then can contribute both inorganic and organic nitrogen to freshwater, coastal, and marine eutrophication, where nitrogen is also typically a limiting nutrient. This is also the case with increased levels of phosphorus. However, because phosphorus is generally much less soluble than nitrogen, it is leached from the soil at a much slower rate than nitrogen. Consequently, phosphorus is much more important as a limiting nutrient in aquatic systems.

Ecological Effects

Eutrophication was recognized as a water pollution problem in European and North American lakes and reservoirs in the mid-20th century. Since then, it has become more widespread. Surveys showed that 54% of lakes in Asia are eutrophic; in Europe, 53%; in North America, 48%; in South America, 41%; and in Africa, 28%.

Many ecological effects can arise from stimulating primary production, but there are three particularly troubling ecological impacts: decreased biodiversity, changes in species composition and dominance, and toxicity effects.

- Increased biomass of phytoplankton
- Toxic or inedible phytoplankton species
- Increases in blooms of gelatinous zooplankton
- Increased biomass of benthic and epiphytic algae
- Changes in macrophyte species composition and biomass
- Decreases in water transparency (increased turbidity)
- Colour, smell, and water treatment problems
- Dissolved oxygen depletion
- Increased incidences of fish kills
- Loss of desirable fish species
- Reductions in harvestable fish and shellfish
- Decreases in perceived aesthetic value of the water body

Decreased Biodiversity

When an ecosystem experiences an increase in nutrients, primary producers reap the benefits first. In aquatic ecosystems, species such as algae experience a population increase (called an algal bloom). Algal blooms limit the sunlight available to bottom-dwelling organisms and cause wide swings in the amount of dissolved oxygen in the water. Oxygen is required by all aerobically respiring plants and animals and it is replenished in daylight by photosynthesizing plants and algae.

Under eutrophic conditions, dissolved oxygen greatly increases during the day, but is greatly reduced after dark by the respiring algae and by microorganisms that feed on the increasing mass of dead algae. When dissolved oxygen levels decline to hypoxic levels, fish and other marine animals suffocate.

As a result, creatures such as fish, shrimp, and especially immobile bottom dwellers die off. In extreme cases, anaerobic conditions ensue, promoting growth of bacteria such as *Clostridium botulinum* that produces toxins deadly to birds and mammals. Zones where this occurs are known as dead zones.

New Species Invasion

Eutrophication may cause competitive release by making abundant a normally limiting nutrient. This process causes shifts in the species composition of ecosystems. For instance, an increase in nitrogen might allow new, competitive species to invade and out-compete original inhabitant species. This has been shown to occur in New England salt marshes. In Europe and Asia, the Common carp frequently lives in naturally Eutrophic or Hypereutrophic areas, and is adapted to living in such conditions. The eutrophication of areas outside its natural range partially explain the fish's success in colonising these areas after being introduced.

Toxicity

Some algal blooms, otherwise called "nuisance algae" or "harmful algal blooms", are toxic to plants and animals. Toxic compounds they produce can make their way up the food chain, resulting in animal mortality. Freshwater algal blooms can pose a threat to livestock. When the algae die or are eaten, neuro- and hepatotoxins are released which can kill animals and may pose a threat to humans. An example of algal toxins working their way into humans is the case of shellfish poisoning. Biotoxins created during algal blooms are taken up by shellfish (mussels, oysters), leading to these human foods acquiring the toxicity and poisoning humans. Examples include paralytic, neurotoxic, and diarrhoetic shellfish poisoning. Other marine animals can be vectors for such toxins, as in the case of ciguatera, where it is typically a predator fish that accumulates the toxin and then poisons humans.

Sources of High Nutrient Runoff

In order to gauge how to best prevent eutrophication from occurring, specific sources that contribute to nutrient loading must be identified. There are two common sources of nutrients and organic matter: point and nonpoint sources.

Point Sources

Point sources are directly attributable to one influence. In point sources the nutrient waste travels directly from source to water. Point sources are relatively easy to regulate.

Nonpoint Sources

Nonpoint source pollution (also known as 'diffuse' or 'runoff' pollution) is that which comes from ill-defined and diffuse sources. Nonpoint sources are difficult to regulate and usually vary spatially and temporally (with season, precipitation, and other irregular events).

It has been shown that nitrogen transport is correlated with various indices of human activity in watersheds, including the amount of development. Ploughing in agriculture and development are activities that contribute most to nutrient loading. There are three reasons that nonpoint sources are especially troublesome:

Soil Retention

Nutrients from human activities tend to accumulate in soils and remain there for years. It has been shown that the amount of phosphorus lost to surface waters increases linearly with the amount of phosphorus in the soil. Thus much of the nutrient loading in soil eventually makes its way to water. Nitrogen, similarly, has a turnover time of decades.

Runoff to Surface Water and Leaching to Groundwater

Nutrients from human activities tend to travel from land to either surface or ground water. Nitrogen in particular is removed through storm drains, sewage pipes, and other forms of surface runoff. Nutrient losses in runoff and leachate are often associated with agriculture. Modern agriculture often involves the application of nutrients onto fields in order to maximise production. However, farmers frequently apply more nutrients than are taken up by crops or pastures. Regulations aimed at minimising nutrient exports from agriculture are typically far less stringent than those placed on sewage treatment plants and other point source polluters. It should be also noted that lakes within forested land are also under surface runoff influences.

Runoff can wash out the mineral nitrogen and phosphorus from detritus and in consequence supply the water bodies leading to slow, natural eutrophication.

Atmospheric Deposition

Nitrogen is released into the air because of ammonia volatilization and nitrous oxide production. The combustion of fossil fuels is a large human-initiated contributor to atmospheric nitrogen pollution. Atmospheric deposition (e.g., in the form of acid rain) can also affect nutrient concentration in water, especially in highly industrialized regions.

Other Causes

Any factor that causes increased nutrient concentrations can potentially lead to eutrophication. In modelling eutrophication, the rate of water renewal plays a critical role; stagnant water is allowed to collect more nutrients than bodies with replenished water supplies. It has also been shown that the drying of wetlands causes an increase in nutrient concentration and subsequent eutrophication blooms.

Prevention and Reversal

Eutrophication poses a problem not only to ecosystems, but to humans as well. Reducing eutrophication should be a key concern when considering future policy, and a sustainable solution for everyone, including farmers and ranchers, seems feasible. While eutrophication does pose problems, humans should be aware that natural runoff (which causes algal blooms in the wild) is common in ecosystems and should thus not reverse nutrient concentrations beyond normal levels. Cleanup measures have been mostly, but not completely, successful. Finnish phosphorus removal measures started in the mid-1970s and have targeted rivers and lakes polluted by industrial and municipal discharges. These efforts have had a 90% removal efficiency. Still, some targeted point sources did not show a decrease in runoff despite reduction efforts.

Shellfish in Estuaries: Unique Solutions

One proposed solution to eutrophication in estuaries is to restore shellfish populations, such as oysters and mussels. Oyster reefs remove nitrogen from the water column and filter out suspended solids, subsequently reducing the likelihood or extent of harmful algal blooms or anoxic conditions. Filter feeding activity is considered beneficial to water quality by controlling phytoplankton density and sequestering nutrients, which can be removed from the system through shellfish harvest, buried in the sediments, or lost through denitrification. Foundational work toward the idea of improving marine water quality through shellfish cultivation was conducted by Odd Lindahl et al., using mussels in Sweden. In the United States, shellfish restoration projects have been conducted on the East, West and Gulf coasts.

Minimizing Nonpoint Pollution: Future Work

Nonpoint pollution is the most difficult source of nutrients to manage. The literature suggests, though, that when these sources are controlled, eutrophication decreases. The following steps are recommended to minimize the amount of pollution that can enter aquatic ecosystems from ambiguous sources.

Riparian Buffer Zones

Studies show that intercepting non-point pollution between the source and the water is a successful means of prevention. Riparian buffer zones are interfaces between a flowing body of water and land, and have been created near waterways in an attempt to filter pollutants; sediment and nutrients are deposited here instead of in water. Creating buffer zones near farms and roads is another possible way to prevent nutrients from travelling too far. Still, studies have shown that the effects of atmospheric nitrogen pollution can reach far past the buffer zone. This suggests that the most effective means of prevention is from the primary source.

Prevention Policy

Laws regulating the discharge and treatment of sewage have led to dramatic nutrient reductions to surrounding ecosystems, but it is generally agreed that a policy regulating agricultural use of fertilizer and animal waste must be imposed. In Japan the amount of nitrogen produced by livestock is adequate to serve the fertilizer needs for the agriculture industry. Thus, it is not unreasonable to command livestock owners to clean up animal waste—which when left stagnant will leach into ground water.

Policy concerning the prevention and reduction of eutrophication can be broken down into four sectors: Technologies, public participation, economic instruments, and cooperation. The term technology is used loosely, referring to a more widespread use of existing methods rather than an appropriation of new technologies. As mentioned before, nonpoint sources of pollution are the primary contributors to eutrophication, and their effects can be easily minimized through common agricultural practices. Reducing the amount of pollutants that reach a watershed can be achieved through the protection of its forest cover, reducing the amount of erosion leeching into a watershed. Also, through the efficient, controlled use of land using sustainable agricultural practices to minimize land degradation, the amount of soil runoff and nitrogen-based fertilizers reaching a watershed can be reduced. Waste disposal technology constitutes another factor in eutrophication prevention.

Because a major contributor to the nonpoint source nutrient loading of water bodies is untreated domestic sewage, it is necessary to provide treatment facilities to highly urbanized areas, particularly those in underdeveloped nations, in which treatment of domestic waste water is a scarcity. The technology to safely and efficiently reuse waste water, both from domestic and industrial sources, should be a primary concern for policy regarding eutrophication.

The role of the public is a major factor for the effective prevention of eutrophication. In order for a policy to have any effect, the public must be aware of their contribution to the problem, and ways in which they can reduce their effects.

Programmes instituted to promote participation in the recycling and elimination of wastes, as well as education on the issue of rational water use are necessary to protect water quality within urbanized areas and adjacent water bodies.

Economic instruments, "which include, among others, property rights, water markets, fiscal and financial instruments, charge systems and liability systems, are gradually becoming a substantive component of the management tool set used for pollution control and water allocation decisions." Incentives for those who practice clean, renewable, water management technologies are an effective means of encouraging pollution prevention. By internalizing the costs associated with the negative effects on the environment, governments are able to encourage a cleaner water management.

Because a body of water can have an effect on a range of people reaching far beyond that of the watershed, cooperation between different organizations is necessary to prevent the intrusion of contaminants that can lead to eutrophication. Agencies ranging from state governments to those of water resource management and non-governmental organizations, going as low as the local population, are responsible for preventing eutrophication of water bodies. In the United States, the most well known inter-state effort to prevent eutrophication is the Chesapeake Bay.

Nitrogen Testing and Modelling

Soil Nitrogen Testing (N-Testing) is a technique that helps farmers optimize the amount of fertilizer applied to crops. By testing fields with this method, farmers saw a decrease in fertilizer application costs, a decrease in nitrogen lost to surrounding sources, or both. By testing the soil and modelling the bare minimum amount of fertilizer needed, farmers reap economic benefits while reducing pollution.

Organic Farming

There has been a study that found that organically fertilized fields "significantly reduce harmful nitrate leaching" over conventionally fertilized fields. However, a more recent study found that eutrophication impacts are in some cases higher from organic production than they are from conventional production.

Cultural Eutrophication

Cultural eutrophication is the process that speeds up natural eutrophication because of human activity. Due to clearing of land and building of towns and cities, land runoff is accelerated and more nutrients such as phosphates and nitrate are supplied to lakes and rivers, and then to coastal estuaries and bays. Extra nutrients are also supplied by treatment plants, golf courses, fertilizers, and farms.

These nutrients result in an excessive growth of plant life known as an algal bloom. This can change a lake's natural food web, and also reduce the amount of dissolved oxygen in the water for organisms to breathe. Both these things cause animal and plant death rates to increase as the plants take in poisonous water while the animals drink the poisoned water. This contaminates water, making it undrinkable, and sediment quickly fills the lake. Cultural eutrophication is a form of water pollution.

Cultural eutrophication also occurs when excessive fertilizers run into lakes and rivers. This encourages the growth of algae (algal bloom) and other aquatic plants. Following this, overcrowding occurs and plants compete for sunlight, space and oxygen. Overgrowth of water plants also blocks sunlight and oxygen for aquatic life in the water, which in turn threatens their survival. Algae also grows easily, thus threatening other water plants no matter whether they are floating, half-submerged, or fully submerged. Not only does this cause algal blooming, it can cause an array of more long term effects on the water such as damage to coral reefs and deep sea animal life. It also speeds up the damage of both marine and also affects humans if the effects of algal blooming is too drastic. Fish will die and there will be lack of food in the area. Nutrient pollution is a major cause of algal blooming, and should be minimized.

The Experimental Lakes Area (ELA), Ontario, Canada is a fully equipped, year-round, permanent field station that uses the whole ecosystem approach and long-term, whole-lake investigations of freshwater focusing on cultural eutrophication. ELA is currently cosponsored by the Canadian Departments of Environment and Fisheries and Oceans, with a mandate to investigate the aquatic effects of a wide variety of stresses on lakes and their catchments.

Fisheries of River

Rivers have formed nuclei for human settlement from the origins of mankind. Many of the earliest civilizations emerged upon the fertile floodplains and since about 5000 b.p., when the earliest systematic colonization of the Nile, Mesopotamic, Indus and Chinese rivers occurred there has been a concentrated effort aimed at the domination of the hydro-logical regimes for the benefit of agriculture. The Roman culture and later that of Western Europe impounded many smaller rivers for water power. In many of the more arid parts of the world streams and rivers were manipulated to provide water for irrigation. These trends have increased until the present day efforts at impoundment, deviation and canalization have left few rivers with undisturbed channels. To the environmental impacts of hydraulic engineering must be added the hazards of the contamination of the waters with a variety of agricultural, domestic and industrial chemicals. In addition bad, or non- existent basin management, deforestation and farming of marginal hill slope lands has increased erosion and the silt loads of rivers resulting in rapid modification of the lowland reaches of the river. These changes not only modify the environment, depriving the fish of living space and access to parts of the river necessary for the completion of their life cycles, but also changes in the quality and quantity of the water in which they live.

The fish communities of rivers have provided the basis for fisheries presumably from the earliest phases of human occupation of river valleys.

In general studies of fish communities have lagged behind those of lakes and reservoirs although there has been an increase in interest in this topic in the last decade. Practical concern with the management of rivers for fisheries began towards the end of the last century in North America and Europe and led to research in such waters in support of stocking and physical improvement programmes principally in support of srort fisheries for salmonids. These led to early classifications of rivers into zones in Eastern Europe where commercial fisheries for coarse fish were also economically important.

> *Rivers drain all but the most arid areas of the earth through channels that are regulated by physical laws that impose on them certain forms. The ideal form is rarely encountered in practice and represents an end point to which geographic process tend. In general a river may be divided into two principal zones, the steep and fast flowing rhithron upstream and the sluggish and flat potamon downstream. While conditions in an individual system are highly variable along its length, similar reaches of different rivers differ much less even between continents and at different latitudes. All continents have a series of major river systems which consist not only of the river channels but also the swamps, lakes and seasonally flooded lands associated with them.*
>
> *Most rivers are highly conditioned by the patterns of precipitation in their basins. Differences in rainfall intensity throughout the year generate a flood wave that progresses downstream in the majority of rivers (flood rivers), although singular geographic circumstances may distribute discharge more evenly throughout the year in some systems (reservoir rivers). The number of reservoir rivers is increasing through flow regulation and dam building. Although the basic nature of the river is determined by the rocks over which it flows, the flood regime seasonally modifies the physical and chemical conditions within the river particularly in the tropics. In higher latitudes other features of climate, such as insolation or air temperature exert an increasing influence.*
>
> *Seasonal changes in discharge, nutrient concentrations, pH, temperature and dissolved oxygen in their turn influence the composition and abundance of the plant and animal communities inhabiting the river. These changes are particularly marked in the floodplain area of the potamon where the rise in water during the floods inundates extensive areas of land flanking the main channels. This increase in living space, together with*

the release of nutrients associated with the submersion of the soil produces an annual surge of primary productivity closely followed by an expansion in biomass of animal communities.

The number of fish species inhabiting rivers is a function of the size of the river, with larger basins such as the Amazon having well over 1000 species. The individual species are highly adapted to the conditions in the type of river reach in which they live. Such adaptations are not only morphological but also behavioural and some species have developed extensive migrations to avoid adverse conditions or for breeding and feeding. Alternatively elaborate breeding mechanisms have also evolved.

Other features of the biology of fishes are linked to the hydrological cycles within the river. Thus, the flood is associated with spawning in the majority of species when the abundance of living space and food provides the best conditions for the survival and growth of the young fish. Such is the influence of these factors that in years of more intense flooding survival and growth are so improved that the total biomass of the fish community rises and a strong year class is produced for transmission on to other years. In reservoir rivers and in the rhithron seasonal and year-to-year differences of this type are not so marked.

Fish communities in rivers provide the basis for fisheries which are pursued with a great variety of gear. Fishing intensity is also seasonal and is tied either to variations in temperature or to the flood. Because the fish community can vary in abundance with the fluctuations in flood strength catch is similarly correlated with years of high catch following after years of particularly good flooding. Fish communities respond to increases in fishing pressure in a number of ways. In general catch in rivers having simple fish communities follow a typical yield curve whereas those in which communities are complex show a plateau in catch which may persist over a great range of effort. This plateau masks changes in the composition of the community with a drift from large, slow growing to small, fast growing forms.

Rivers and their basins are used for many purposes other than fisheries. Many of these maodify the qualit or quantity of water in the system and thus interact with the fish communities in the river to their detriment. Management of the river for fisheries therefore becomes increasingly important as the intensity of use of the river rises. Fisheries themselves also require management which may be accomplished either by direct interventions on the fish stock or by legislative or economic activities on the fishermen themselves. As the river becomes increasingly modified the capture fisheries originally pursued there become less viable. Although the water courses may continue to provide food or recreation the major developmental emphasis is towards replacement activities such as aquaculture on the former floodplain or the creation of new fisheries in the reservoirs associated with the main channel.

Riverine Fisheries in India

The riverine fisheries are a part of inland fisheries where the fishes are captured directly from the different river systems with the help of scientifically made fish crafts and gears. The riverine fishery resources in India is immense as large number of productive riverine system is present in this country.

The river system in India mainly constitute five river systems, namely the Ganges, the Brahmaputra, the Indus, the peninsular East coast (comprising the Mahanandi, the Godavari, the Krishna and the Cauvery) and the West Coast (including west of western ghat, basins of Narmada and Tapti). .

The Ganga River System: It is the largest river system in India. The Ganges is a perennial river, originating from the Gangotri near Himalayas, enters the plains at Hardwar, passes through the states of U.P., Bihar and West Bengal and ultimately joins the Bay of Bengal. The principal tributaries are rivers Ramganga, Gomati, Ghaghra, Gandak, Kosi, Yamuna (with Chambal, Betwa and Keno) and Tons and Sons. All this amounts a catchment area of about 9.72 lakhs sq. km.

The major carps of this river system are Cirrhinus mrigala, Labeo rohita, Labeo calbasu, Mystus oar, Mystus seenghala, Wallago attu, Hilsa ilisha and several other miscellaneous species.

The Brahmaputra River System: This river system originates from glacier mass near Mansarower Lake, crosses through Tibet (where it is called Ptsango), Arunachal, Assam, and Bangladesh joins the Ganga at Goalundo. The Brahmputra and the Ganga after their merger is laiown as Padma which joins the Bay of Bengal through Meghna estuary. There is large number of its tributaries in Assam and this river system cover a total catchment area of about 2,900 sq. km. the amount of commercial production from this river system is not well known, however, the important fishes of this system are Labeo gonius, Wallago attu, Puntius sarana, Notopterus notopterus. Notopterus chitala, and Mystus seenhgala.

The Indus River System: The Indus rises about 100 Km. North of Mansarowar and flows North West through Tibet before entering Kashmir. After flowing for about 800 kg. it turns south through the Ladakh range. It enters Pakistan through Kashmir. It has five tributaries on its le t bank, namely, the Jhelum, Chenab, Ravi, Beas and Sutlej. It has not been possible to estimate the amount of commercial production from this river system. The fishes are of mixed type.

The East Coast System: It constitutes of four major rivers. Godavari originates from Deolali hills in Northern Western Ghat and terminates in Bay of Bengal. It has a total catchment area of 315 980 sq km. The river Krishna originates in the western ghat, ranges south of Poona to its delta on the east coast. It has a catchment area of about 233,229 sq. km. River Cauvery originates from the Brahmagin Hills on the western ghat, flows in south easterly direction and terminates in Bay of Bengal at Tamil Nadu.

The main fishes of Godavri and Krishna are Labeo fimbriatus, Cirrhinus mrigala, Labeo calbasu, Catla catla, Mystus oar, Mystus seenghala, Silomia childreni, Pangasius pangasius, Wallago attu, Hilsa ilisha etc. Cauvery has some special features in fish fauna. It consists of Acrossocheilus hexagonelopis, Tor putitora, Barbus carnaticus, Barbus dubias, Labeo kontius, Labeo oriza, Osteochilus brevidorsalis, Osteochilus nashi etc.

The Western Coast River System: It has two important rivers namely Narmada and Tapti. The Narmada originates in the Amarkantak Hills, Madhya Pradesh and terminates in the Gulf of Cambay after crossing Gujarat state. The catchment area is about C. 94,235 sq. km. Tor tor forms the bulk of fish produced from this river. Besides this Labeo fimbriatus, Labeo calbasu, cirrhinus mrigala, Puntius sarana, Cirrhinus reba, Rita pavimentata, Mystus seenghala, Clupisoma garua, Ompak bimaculatus, Notopterus notopterus etc.

River Tapti has its origin in Mount Vindhya crosses through Madhya Pradesh, Maharashtra and Gujrat and ultimately join the

Arabian Sea at Dumas near Surat. It has a catchment area of about C. 48,000 sq. km. Fishes are Tor tor, Labeo fimbriatus, Labeo calbasu, Labeo bata, Labeo boggut, Mystus oar, Mystus seenghala etc.

Inland Fishing Crafts and Gears: India is a vast country with large number of river systems. The fishing crafts and gears are not uniformly used throughout the country. Although, scientifically mechanized crafts and gears have been developed, mostly the locally made non-mechanized ones are in practice. Among the crafts several types of crafts like Dogas, Buffaloskin, Bananna stem, stem, Bamboos, etc., are used. In stagnant water "dongas" are best suited. They are made by hollowing out the stem of Palmyrapalm which become an efficient craft for floating. They are more commonly used in West Bengal.

"Vallam" are similar type of craft but are more strongly build to face the estuaries water and are commonly used in Kerala. In river Ganga inflated buffalo skins tied together, or bamboo plate-forms supported by earthen pots are generally used. In West Bengal and Bihar small boats known as "Dinghis" are used for the operation of small nets. The larger types of boat for operation of larger nets are also used which are commonly called as "Chandi Nauka".

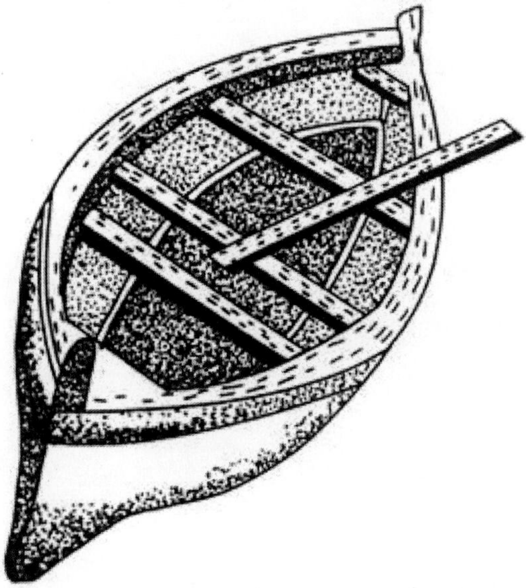

As different productive areas are of diverse nature, the gears used and the method of fish capturing also varies from place to place. At several places especially in small reservoirs ponds, spears, bows and arrows, rod and line are generally used by individual persons. In larger ponds and lakes different size of seine and driftnets' are used.

Large seines called as Mahajals are generally used in the Ganges. For fishing in small areas "dragnets are best suited.

Larger form called as Berjal are extensively used in the Ganges. Berjal are without sinkers and are spreader in the river in a semicircular fashion. For carrying and spreading the net boats are used. The two ends of the net are held by several persons on the shore for dragging the net. Rangoon net with large number of small sized sinkers tied to the foot row is used in the areas with non uniform and uneven bottom.

Specific type of nets is used for specific type of fishes. For Hilsa fishing Kona jal or Bhasa gulli having conical pockets are used in West Bengal and Bihar. Katla Jal is used specially tî capture carps and is common in Bihar and Uttar Pradesh. For cat fishes "Hela Jal" which is a dip net consisting of two light bamboo poles crossing each other are in practice. In river Narmada a special type of dip net known as "Khoila Jal" are used for the capture of grey mullet. Small nets for individual use having a wide mouth and narrow end are being used in pond fishing. Besides the above mentioned nets and gears, several other types are also in practice, detail coverage of which is not possible here.

Chapter 2

Natural Resources: Wetlands their Classification Functions and Values

A wetland is a land area that is saturated with water, either permanently or seasonally, such that it takes on the characteristics of a distinct ecosystem.

Primarily, the factor that distinguishes wetlands from other land forms or water bodies is the characteristic vegetation of aquatic plants, adapted to its unique hydric soil. Wetlands play a number of roles in the environment, principally water purification, flood control, and shoreline stability. Wetlands are also considered the most biologically diverse of all ecosystems, serving as home to a wide range of plant and animal life.

Wetlands occur naturally on every continent except Antarctica, the largest including the Amazon River basin, the West Siberian Plain, and the Pantanal. The water found in wetlands can be freshwater, brackish, or saltwater.

The main wetland types include swamps, marshes, bogs, and fens; and sub-types include mangrove, carr, pocosin, and varzea. They can also be constructed artificially as a water management tool, which may play a role in the developing field of water-sensitive urban design.

The UN Millennium Ecosystem Assessment determined that environmental degradation is more prominent within wetland systems than any other ecosystem on Earth. International conservation efforts are being used in conjunction with the development of rapid assessment tools to inform people about wetland issues.

Ecology of Wetland

Many ecologically and economically important species call wetlands home for at least part of their lives. For instance, commercially important fishes and shellfish, including shrimp, blue crab, oysters, salmon, trout, and seatrout rely on, or are associated with, wetlands. Wetlands are also critical habitat for migratory birds and waterfowl, including ducks, egrets, and geese. While covering only 6% of the Earth's surface, wetlands provide a disproportionately high number of ecosystem services, in addition to maintaining biodiversity. For instance, wetlands also mitigate floods, protect coastal areas from storms, improve water quality, recharge groundwater aquifers, serve as sinks, sources, or transformers of materials, and produce food and goods for human use. When evaluating the economic value of these various functions, Costanza *et al.* concluded that the economic value provided by wetland ecosystems exceeded that provided by lakes, streams, forests, and grasslands and was second only to that provided by coastal estuaries.

Increasing recognition of the value and importance of wetland ecosystems over the last century led to the creation of laws, regulations, and plans to restore and protect wetlands around the world. With mitigation, wetlands are created, restored, or enhanced to offset or replace wetland loss due to development. The Ramsar Convention, an international treaty aimed at conserving wetlands, requires member countries to develop national wetland policies, to establish wetland reserves, and to designate one or more wetlands as an area of international importance. All these efforts are designed to protect or conserve wetlands and the ecosystem services they provide.

For most wetlands, the sources of inflows (e.g., precipitation, surface flow, groundwater flow, tides) and outflows (e.g., evapotranspiration, surface flow, groundwater flow, tides) change over time. As such, hydrology is rarely stable but fluctuates over time resulting in pulsing hydroperiods. Hydrologic pulses can alter productivity along a flooding gradient by altering the extent of flood subsidies and stresses in a wetland. When subsidies are high but stress is relatively low, pulses can promote productivity by introducing water, sediments, and nutrients while also removing waste materials and toxins.

Flooding can affect the physiochemistry of wetlands in various ways. Water can introduce or remove sediment, salt, nutrients or other materials from wetlands, thereby influencing its soil and water chemistry. Hydrology also influences the structure and function of wetland ecosystems through its influence on species richness, productivity, rates of organic matter accumulation, and nutrient cycling. Hydrology may restrict species richness in areas subject to long-term flooding while enhancing it in areas with variable or pulsing hydroperiods. Similarly, productivity is typically lower in permanently flooded, stagnant wetlands, or in drained wetlands than in slow-flowing or seasonally flooded wetlands. The anaerobic conditions created under these inundated or flooded conditions often limit decomposition rates, thereby promoting organic matter accumulation in soils, and can alter reduction-oxidation reactions controlling nutrient transformations in wetland soils.

Oxygen Availability

The inundation or saturation of wetland soils by water leads to the formation of anaerobic conditions as oxygen is depleted faster than it can be replaced by diffusion. The rate of oxygen loss in flooded soils can vary depending on other soil conditions, such as temperature and rates of microbial respiration. In most wetlands, small, oxidized layers of soils may persist on the surface or around the roots of vascular plants, but generally, anaerobic, or reduced, conditions prevail.

The prevalence of anaerobic conditions in wetlands has a tremendous impact on their biogeochemistry, with important implications for carbon, nitrogen, phosphorus, iron, manganese, and sulphur transformations. Wetlands can function as sources, sinks, or transformers of these materials, depending on inflows, outflows, and internal cycling rates. One of the most important biogeochemical cycles in wetlands is the nitrogen cycle, and while the potential transformations are not unique to wetlands, the dominance of anaerobic transformations does set wetlands apart from other ecosystems.

One such anaerobic transformation is denitrification, in which nitrate is lost to the atmosphere via conversion to nitrogen gas or nitrous oxide by bacteria. In many wetlands, nutrient availability is dramatically altered by agriculture or other practices that increase nutrient loading, contributing to changes in ecosystem structure and function. Through processes like denitrification and plant uptake, wetlands can help remove some of this excess nitrogen introduced to wetland and aquatic ecosystems.

Wetland Communities and Ecosystems

Because of the predominance of water and anaerobic conditions in wetlands, the organisms living there, especially rooted plants, often exhibit remarkable adaptations to deal with the stresses imposed by flooding. These adaptations, including pressurized gas flow, creation of oxidized root zones, and anaerobic respiration, allow wetland plants to remain productive under otherwise stressful conditions, making wetlands among the most productive ecosystems in the world. This high primary production, in turn, supports high rates of secondary production, rates that can exceed those of terrestrial ecosystems

Abiotic Factors that Influence Wetland Ecosystems

The movement, distribution, and quality of water is the primary factor influencing wetland structure and function. To be classified as a wetland, the presence of water must contribute to the formation of hydric soils, which are formed under flooded or saturated conditions persisting long enough for the development of anaerobic conditions during the growing season. Water conditions in wetlands can vary tremendously with respect to the timing and duration of surface water inundation as well as seasonal patterns of inundation.

In coastal wetlands, tidal influence drives the movement and distribution of water and can range from permanent flooding in subtidal wetlands to less frequent flooding in others, with changes in water level occurring daily or semi-daily. Inland wetlands, which lack daily tidal influences, can also be permanently flooded on one extreme or intermittently flooded on the other extreme, with fluctuations over time often occurring seasonally. It is the balance of water inflows and outflows, or the water budget, as well as the geomorphology and soils that determine the timing, duration, and patterns of flooding in a wetland.

Wetlands exist along soil-moisture gradients, with wetter soils at lower elevations and drier soils at higher elevations. Wetland plant communities develop in response to this environmental gradient based primarily on their individual abilities to tolerate flooding and anaerobic soils but also in response to biotic interactions with other species. Establishment of plant species along an environmental gradient can contribute to sharp plant zonation patterns, as can be seen in coastal wetlands where species separate out along an elevation gradient in response to differences in flooding and salinity.

The development of these productive and often diverse plant communities fuels complex food webs that not only sustain microbial communities through large inputs of detritus to wetland soils but also support diverse communities of animals that utilize wetlands for part or all of their lives. Detritivores, such as shredding insects and crayfish, can utilize dead plant material as their primary energy source, while others (e.g., marsh periwinkle snails) help process organic matter for subsequent use by other organisms. Herbivory of algae by invertebrates and small fish and of plant biomass by some invertebrates, birds, and mammals (e.g., grasshoppers, geese and muskrats) is a significant energy source for primary consumers in many wetlands. Secondary production by these primary consumers supports higher trophic levels, including predatory insects, fishes, reptiles, amphibians, birds, and mammals.

As awareness of wetland ecosystem services and values has increased, wetland ecological research also has increased. Wetland ecologists examine interactions between species and their environment, recognizing the important role that hydrology plays in shaping the physicochemical environment and biological communities in wetlands. Within that framework, ecologists can examine a near-endless array of ecological topics, from the physiology of species coping with flooding stress and anoxia to species interactions, to the impacts of and feedbacks to global climate change.

The diversity of wetland types, the biodiversity they support, and the numerous functions they provide make wetlands an exciting and rewarding arena in which to explore fundamental ecological questions. And it is this science that informs efforts to manage, restore, and conserve the wetlands of the world.

Definitions

A patch of land that develops pools of water after a rain storm would not be considered a "wetland," even though the land is wet. Wetlands have unique characteristics: they are generally distinguished from other water bodies or landforms based on their water level and on the types of plants that live within them. Specifically, wetlands are characterized as having a water table that stands at or near the land surface for a long enough period each year to support aquatic plants.

A more concise definition is a community composed of hydric soil and hydrophytes.

Wetlands have also been described as ecotones, providing a transition between dry land and water bodies. Mitsch and Gosselink write that wetlands exist "...at the interface between truly terrestrial ecosystems and aquatic systems, making them inherently different from each other, yet highly dependent on both."

In environmental decision-making, there are subsets of definitions that are agreed upon to make regulatory and policy decisions.

Technical Definitions

A wetland is "an ecosystem that arises when inundation by water produces soils dominated by anaerobic processes, which, in turn, forces the biota, particularly rooted plants, to adapt to flooding." There are four main kinds of wetlands — marsh, swamp, bog and fen (bogs and fens both being types of mires). Some experts also recognize wet meadows and aquatic ecosystems as additional wetland types. The largest wetlands in the world include the swamp forests of the Amazon and the peatlands of Siberia.

Ramsar Convention definition

Under the Ramsar international wetland conservation treaty, wetlands are defined as follows:

- Article 1.1: "...wetlands are areas of marsh, fen, peatland or water, whether natural or artificial, permanent or temporary, with water that is static or flowing, fresh, brackish or salt, including areas of marine water the depth of which at low tide does not exceed six metres."
- Article 2.1: "[Wetlands] may incorporate riparian and coastal zones adjacent to the wetlands, and islands or bodies of marine water deeper than six metres at low tide lying within the wetlands."

Regional Definitions

Although the general definition given above applies around the world, each county and region tends to have its own definition for legal purposes. In the United States, wetlands are defined as "those areas that are inundated or saturated by surface or groundwater at a frequency and duration sufficient to support, and that under normal circumstances do support, a prevalence of vegetation typically adapted for life in saturated soil conditions. Wetlands generally include swamps, marshes, bogs and similar areas". This definition has been used in the enforcement of the Clean Water Act. Some US states, such as Massachusetts and New York, have separate definitions that may differ from the federal government's.

In the United States Code (16 U.S.C., Section 3801(a)(18)), the term wetland is defined "as land that (A) has a predominance of hydric soils, (B) is inundated or saturated by surface or groundwater at a frequency and duration sufficient to support a prevalence of hydrophytic vegetation typically adapted for life in saturated soil conditions and C) under normal circumstances supports a prevalence of such vegetation." Related to this legal definitions, the term "normal circumstances" are conditions expected to occur during the wet portion of the growing season under normal climatic conditions (not unusually dry or unusually wet), and in the absence of significant disturbance. It is not uncommon for a wetland to be dry for long portions of the growing season. Wetlands can be dry during the dry season and abnormally dry periods during the wet season, but under normal environmental conditions the soils in a wetland will be saturated to the surface or inundated such that the soils become anaerobic, and those conditions will persist through the wet portion of the growing season.

Ecology

The most important factor producing wetlands is flooding. The duration of flooding determines whether the resulting wetland has aquatic, marsh or swamp vegetation. Other important factors include fertility, natural disturbance, competition, herbivory, burial and salinity. When peat accumulates, bogs and swamps arise.

Characteristics

Wetlands vary widely due to local and regional differences in topography, hydrology, vegetation, and other factors, including human involvement. Wetlands can be divided into two main classes: tidal and non-tidal areas.

Hydrology

Wetland hydrology is associated with the spatial and temporal dispersion, flow, and physiochemical attributes of surface and ground water in its reservoirs. Based on hydrology, wetlands can be categorized as riverine (associated with streams), lacustrine (associated with lakes and reservoirs), and palustrine (isolated). Sources of hydrological flows into wetlands are predominately precipitation, surface water, and ground water. Water flows out of wetlands by evapotranspiration, surface runoff, and sub-surface water outflow. Hydrodynamics (the movement of water through and from a wetland) affects hydroperiods (temporal fluctuations in water levels) by controlling the water balance and water storage within a wetland.

Landscape characteristics control wetland hydrology and hydrochemistry. The O_2 and CO_2 concentrations of water depend on temperature and atmospheric pressure. Hydrochemistry within wetlands is determined by the pH, salinity, nutrients, conductivity, soil composition, hardness, and the sources of water. Water chemistry of wetlands varies across landscapes and climatic regions. Wetlands are generally minerotrophic with the exception of bogs.

Bogs receive their water from the atmosphere and therefore their water has low mineral ionic composition because ground water has a higher concentration of dissolved nutrients and minerals in comparison to precipitation.

The water chemistry of fens ranges from low pH and low minerals to alkaline with high accumulation of calcium and magnesium because they acquire their water from precipitation as well as ground water.

Role of Salinity

Salinity has a strong influence on wetland water chemistry, particularly in wetlands along the coast. In non-riverine wetlands, natural salinity is regulated by interactions between ground and surface water, which may be influenced by human activity.

Soil

Carbon is the major nutrient cycled within wetlands. Most nutrients, such as sulphur, phosphorus, carbon, and nitrogen are found within the soil of wetlands. Anaerobic and aerobic respiration in the soil influences the nutrient cycling of carbon, hydrogen, oxygen, and nitrogen, and the solubility of phosphorus thus contributing to the chemical variations in its water. Wetlands with low pH and saline conductivity may reflect the presence of acid sulphates and wetlands with average salinity levels can be heavily influenced by calcium or magnesium. Biogeochemical processes in wetlands are determined by soils with low redox potential.

Biota

The biota of a wetland system includes its vegetation zones and structure as well as animal populations. The most important factor affecting the biota is the duration of flooding. Other important factors include fertility and salinity. In fens, species are highly dependent on water chemistry. The chemistry of water flowing into wetlands depends on the source of water and the geological material in which it flows through as well as the nutrients discharged from organic matter in the soils and plants at higher elevations in slope wetlands. Biota may vary within a wetland due to season or recent flood regimes.

Flora

There are four main groups of hydrophytes that found in wetland systems throughout the world.

Submerged water plants: This type of vegetation is found completely underwater. Submerged wetland vegetation can grow in saline and fresh-water conditions. Some species have underwater flowers, while others have long stems to allow the flowers to reach the surface. Submerged species provide a food source for native fauna, habitat for invertrebrates, and also possess filtration capabilities. Examples include seagrasses and eelgrass.

Floating water plants: Floating vegetation is usually small although it may take up a large surface area in a wetland system. These hydrophytes have small roots and are only found in slow-moving water with rich-nutrient level water Floating aquatic plants are a food resource for avian species. Examples include water lilies, lily pad and duckweed.

Emergent water plants: Emergent water plants can be seen above the surface of the water but whose roots are completely submerged. Many have aerenchyma to transmit oxygen from the atmosphere to their roots. Extensive areas of emergent plants are usually termed marsh. Examples include cattails (Typha) and arrow arum (Peltandra virginica).

Surrounding trees and shrubs: Forested wetlands are generally known as swamps. The upper level of these swamps is determined by high water levels, which are negatively affected by dams. Some swamps can be dominated by a single species, such as silver maple swamps around the Great Lakes. Others, like those of the Amazon Basin, have large numbers of different tree species. Examples include cypress (Taxodium) and mangrove.

Fauna

Fish: Fish are more dependent on wetland ecosystems than any other type of habitat. 75% of the United States' commercial fish and shellfish stocks depend solely on estuaries to survive. Tropical fish species need mangroves for critical hatchery and nursery grounds and the coral reef system for food.

Amphibians: Frogs are the most crucial amphibian species in wetland systems. Frogs need both terrestrial and aquatic habitats in which to reproduce and feed. While tadpoles control algal populations, adult frogs forage on insects. Frogs are used as an indicator of ecosystem health due to their thin skin which absorbs both nutrient and toxins from the surrounding environment resulting in an above average extinction rate in unfavourable and polluted environmental conditions.

Reptiles: Alligators and crocodiles are two common reptilian species. Alligators are found in fresh water along with the fresh water species of the crocodile.

The saltwater crocodile is found in estuaries and mangroves and can be seen in the coastline bordering the Great Barrier Reef in Australia. The Florida Everglades is the only place in the world where both crocodiles and alligators co-exist. Snakes, lizards, goannas, and turtles also can be seen throughout wetlands. Snapping turtles are one of the many kinds of turtles found in wetlands.

Mammals: Multiple small mammals as well as large herbivore and apex species such as the Florida Panther live within and around wetlands. The wetland ecosystem attracts mammals due to its prominent seed sources, invertebrate populations, and numbers of small reptiles and amphibians.

Monotremes: The platypus (*Ornithorhynchus anatinus*) is found in eastern Australia living in freshwater rivers or lakes, and much like the beaver creates dams, create burrows for shelter and protection. The platypus swims through the use of webbed feet.

Platypuses feed on insect larvae, worms, or other freshwater insects hunting mainly by night by the use of their bill. They turn up mud on the bottom of the lake or river, and with the help of the electroreceptors located on the bill, unearth insects and freshwater insects.

The platypus stores their findings in special pouches behind their bill and consumes its prey upon returning to the surface.

Insects and invertebrates: These species total more than half of the 100,000 known animal species in wetlands. Insects and invertebrates can be submerged in the water or soil, on the surface, and in the atmosphere.

Algae

Algae are diverse water plants that can vary in size, colour, and shape. Algae occur naturally in habitats such as inland lakes, intertidal zones, and damp soil and provide a dedicated food source for animals, fish, and invertebrates. There are three main groups of algae:

Plankton are algae which are microscopic, free-floating algae. This algae is so tiny that on average, if fifty of these microscopic algae were lined up end-to-end, it would only measure one millimetre.

Plankton are the basis of the food web and are responsible for primary production in the ocean using photosynthesis to make food. Filamentous algae are long strands of algae cells that form floating mats. Chara and Nitella algae are upright algae that look like a submerged plant with roots.

Climates

Temperature

Temperatures vary greatly depending on the location of the wetland. Many of the world's wetlands are in temperate zones (midway between the North or South Pole and the equator). In these zones, summers are warm and winters are cold, but temperatures are not extreme.

However, wetlands found in the tropic zone, around the equator, are warm year round. Wetlands on the Arabian Peninsula, for example, can reach 50 °C (122 °F) and would therefore be subject to rapid evaporation. In northeastern Siberia, which has a polar climate, wetland temperatures can be as low as –50 °C (–58 °F). In a moderate zone, such as the Gulf of Mexico, a typical temperature might be 11 °C (51 °F). Wetlands are also located in every climatic zone.

Rainfall

The amount of rainfall a wetland receives varies widely according to its area. Wetlands in Wales, Scotland, and Western Ireland typically receive about 1500 mm (or 60 in) per year. In some places in Southeast Asia, where heavy rains occur, they can receive up to 10,000 mm (about 200 in). In the northern areas of North America, wetlands exist where as little as 180 mm (7 inches) of rain fall each year.

Temporal variation:

- Perennial systems
- Seasonal systems
- Episodic (periodic or intermittent) system of the down
- Surface flow may occur in some segments, with subsurface flow in other segments
- Ephemeral (short-lived) systems
- Migratory species

Human-disturbance:

- Encroachment
  - o Drainage
  - o Development
  - o Over-grazing
  - o Mining
  - o Unsustainable water use

- Ecosystem Stress
    - o Water Scarcity
    - o Endangered species
    - o Disruption of breeding grounds
    - o Imbalance in sediment load and nutrient filtration

Peatswamps of South-east Asia

In Southeast Asia, peatswamp forests and soils are being drained, burnt, mined, and overgrazed, contributing severely to climate change. As a result of peat drainage, the organic carbon that was built up over thousands of years and is normally under water is suddenly exposed to the air. It decomposes and turns into carbon dioxide (CO_2), which is released into the atmosphere. Peat fires cause the same process to occur and in addition create enormous clouds of smoke that cross international borders, such as happens every year in Southeast Asia. While peatlands constitute only 3% of the world's land area, their degradation produces 7% of all fossil fuel CO_2 emissions.

Through the building of dams, Wetlands International is halting the drainage of peatlands in Southeast Asia, hoping to mitigate CO_2 emissions. Concurrent wetland restoration techniques include reforestation with native tree species as well as the formation of community fire brigades. This sustainable approach can be seen in Central Kalimantan and Sumatra, Indonesia.

Aquaculture

Concerns are developing over certain aspects of farm fishing, which uses natural waterways to harvest fish for human consumption and pharmaceuticals. This practice has become especially popular in Asia and the South Pacific. Its impact upon much larger waterways downstream has negatively affected many small island developing states.

The function of natural wetlands can be classified by their ecosystem benefits. United Nations Millennium Ecosystem Assessment and Ramsar Convention found wetlands to be of biosphere significance and societal importance in the following areas:

- Flood control
- Groundwater replenishment
- Shoreline stabilisation and storm protection
- Water purification
- Reservoirs of biodiversity

- Wetland products
- Cultural values
- Recreation and tourism
- Climate change mitigation and adaptation

The economic worth of the ecosystem services provided to society by intact, naturally functioning wetlands is frequently much greater than the perceived benefits of converting them to 'more valuable' intensive land use – particularly as the profits from unsustainable use often go to relatively few individuals or corporations, rather than being shared by society as a whole.*-Ramsar convention*

Unless otherwise cited, Ecosystem services is based on the following series of references.

Flood control

Major Wetland Type: Floodplain

Storage Reservoirs and Flood Protection: The wetland system of floodplains is formed from major rivers downstream from their headwaters. Notable river systems that produce large spans of floodplain include the Nile River, the Niger river inland delta, [the Zambezi River flood plain], [the Okavango River inland delta], [the Kafue River flood plain] [the Lake Bangweulu flood plain] (Africa), Mississippi River (USA), Amazon River (South America), Yangtze River (China), Danube River (Central Europe) and Murray-Darling River (Australia). "The floodplains of major rivers act as natural storage reservoirs, enabling excess water to spread out over a wide area, which reduces its depth and speed. Wetlands close to the headwaters of streams and rivers can slow down rainwater runoff and spring snowmelt so that it doesn't run straight off the land into water courses. This can help prevent sudden, damaging floods downstream."

Human-Impact: Converting wetlands through drainage and development have contributed to the issue of irregular flood control through forced adaption of water channels to narrower corridors due to loss of wetland area. These new channels must manage the same amount of precipitation causing flood peaks to be [higher or deeper] and floodwaters to travel faster.

Water management engineering developments in the past century have degraded these wetlands through the construction on artificial embankments. These constructions may be classified as dykes, bunds, levees, weirs, barrages and dams but serve the single purpose of concentrating water into a select source or area. Wetland water sources that were once spread slowly over a large, shallow area are pooled into deep, concentrated locations. Loss of wetland floodplains results in more severe and damaging flooding. Catastrophic human impact in the Mississippi River floodplains was seen in death of several hundred individuals during a levee breach in New Orleans caused by Hurricane Katrina. Ecological catastrophic events from human-made embankments have been noticed along the Yangtze River floodplains after the where the middle of the river has become prone to more frequent and damaging flooding including the loss of riparian vegetation, a 30% loss of the vegetation cover throughout the river's basin, a doubling of the percentage of the land affected by soil erosion, and a reduction in reservoir capacity through siltation build-up in floodplain lakes.

Groundwater Replenishment

Major wetland type: marsh, swamp, & subterranean karst and cave hydrological systems: The surface water which is the water visibly seen in wetland systems only represents a portion of the overall water cycle which also includes atmospheric water and groundwater. Wetland systems are directly linked to groundwater and a crucial regulator of both the quantity and quality of water found below the ground. Wetland systems that are made of permeable sediments like limestone or occur in areas with highly variable and fluctuating water tables especially have a role in groundwater replenishment or water recharge. Sediments that are porous allow water to filter down through the soil and overlying rock into aquifers which are the source of 95% of the world's drinking water. Wetlands can also act as recharge areas when the surrounding water table is low and as a discharge zone when it is too high. Karst (cave) systems are a unique example of this system and are a connection of underground rivers influenced by rain and other forms of precipitation. These wetland systems are capable of regulating changes in the water table on upwards of 130 metres (430 ft).

Human-Impact: Groundwater is an important source of water for drinking and irrigation of crops. Over 1 billion people in Asia and 65% of the public water sources in Europe source 100% of their water from groundwater. Irrigation is a massive use of groundwater with 80% of the world's groundwater used for agricultural production.

Unsustainable abstraction of groundwater has become a major concern. In the Commonwealth of Australia, water licensing is being implemented to control use of the water in major agricultural regions. On a global scale, groundwater deficits and water scarcity is one of the most pressing concerns facing the 21st century.

Shoreline Stabilisation and Storm Protection

Wetland type: Mangroves, Coral Reefs, Saltmarsh: Tidal and inter-tidal wetland systems protect and stabilize coastal zones. Coral reefs provide a protective barrier to coastal shoreline. Mangroves stabilize the coastal zone from the interior and will migrate with the shoreline to remain adjacent to the boundary of the water. The main conservation benefit these systems have against storms and tidal waves is the ability to reduce the speed and height of waves and floodwaters.

Human-Impact: The sheer number of people who live and work near the coast is expected to grow immensely over the next 50 years. From an estimated 200 million people that currently live in low-lying coastal regions, the development of urban coastal centres is projected to increase the population by 5 fold within 50 years.

The United Kingdom has begun the concept of managed coastal realignment. This management technique provides shoreline protection through restoration of natural wetlands rather than through applied engineering. In East Asia, reclamation of coastal wetlands has resulted in widespread transformation of the coastal zone, and up to 65% of coastal wetlands have been destroyed by coastal development.

Water Purification

Wetland Type: Floodplain, Mudflat, Saltmarsh, Mangroves

Nutrient Retention: Wetlands cycle both sediments and nutrients balancing terrestrial and aquatic ecosystems. A natural function of wetland vegetation is the up-take and storage of nutrients found in the surrounding soil and water.

These nutrients are retained in the system until the plant dies or is harvested by animals or humans. Wetland vegetation productivity is linked to the climate, wetland type, and nutrient availability. The grasses of fertile floodplains such as the Nile produce the highest yield including plants such as *Arundo donax* (giant reed), *Cyperus papyrus* (papyrus), *Phragmites* (reed) and *Typha* (cattail, bulrush).

Sediment Traps: Rainfall run-off is responsible for moving sediment through waterways. These sediments move towards larger and more sizable waterways through a natural process that moves water towards oceans. All types of sediments which may be composed of clay, sand, silt, and rock can be carried into wetland systems through this process. Reedbeds or forests located in wetlands act as physical barriers to slow waterflow and trap sediment.

Water purification: Many wetland systems possess biofilters, hydrophytes, and organisms that in addition to nutrient up-take abilities have the capacity to remove toxic substances that have come from pesticides, industrial discharges, and mining activities.

The up-take occurs through most parts of the plant including the stems, roots, and leaves. Floating plants can absorb and filter heavy metals. *Eichhornia crassipes* (water hyacinth), *Lemna* (duckweed) and *Azolla* (water fern) store iron and copper commonly found in wastewater. Many fast-growing plants rooted in the soils of wetlands such as *Typha* (cattail) and *Phragmites* (reed) also aid in the role of heavy metal up-take. Animals such as the oyster can filter more than 200 liters (53 gallons) of water per day while grazing for food, removing nutrients, suspended sediments, and chemical contaminants in the process.

Capacity: The ability of wetland systems to store nutrients and trap sediment is highly efficient and effective but each system has a threshold. An overabundance of nutrient input from fertilizer run-off, sewage effluent, or non-point pollution will cause eutrophication. Upstream erosion from deforestation can overwhelm wetlands making them shrink in size and see dramatic biodiversity loss through excessive sedimentation load. The capacity of wetland vegetation to store heavy metals is affected by waterflow, number of hectares (acres), climate, and type of plant.

Human-Impact: Introduced hydrophytes in different wetland systems can have devastating results. The introduction of water hyacinth, a native plant of South America into Lake Victoria in East Africa as well as duckweed into non-native areas of Queensland, Australia, have overtaken entire wetland systems suffocating the ecosystem due to their phenomenal growth rate and ability to float and grow on the surface of the water.

Constructed Wetlands

The function of most natural wetland systems is not to manage to wastewater, however, their high potential for the filtering and the treatment of pollutants has been recognized by environmental engineers that specialize in the area of wastewater treatment. These constructed artificial wetland systems are highly controlled environments that intend to mimic the occurrences of soil, flora, and microorganisms in natural wetlands to aid in treating wastewater effluent. Artificial wetlands provide the ability to experiment with flow regimes, micro-biotic composition, and flora in order to produce the most efficient treatment process. Other advantages are the control of retention times and hydraulic channels. The most important factors of constructed wetlands are the water flow processes combined with plant growth. Constructed wetland systems can be surface flow systems with only free-floating macrophytes, floating-leaved macrophytes, or submerged macrophytes; however, typical free water surface systems are usually constructed with emergent macrophytes. Constructed wetlands can be adapted to treat raw sewage, secondary domestic sludge, enhance water quality of oxidation ponds' discharge, storm waters, mining waste, and industrial and agricultural waste effluents. The Urrbrae Wetland in Australia was constructed for urban flood control and environmental education. International wastewater management programmes can be seen from Kolkata (Calcutta), India to Arcata, California, USA.

Kolkata's constructed wetland: Kolkata is an example of how constructed wetlands are being utilized in developing countries.

Using the purification capacity of wetlands, the Indian city of Kolkata (Calcutta) has pioneered a system of sewage disposal that is both efficient and environmentally friendly. Built to house one million people, Kolkata is now home to over 10 million, many living in slums. But the 8,000-hectare East Kolkata Wetlands Ramsar Site, a patchwork of tree-fringed canals, vegetable plots, rice paddies and fish ponds – and the 20,000 people that work in them – daily transform one-third of the city's sewage and most of its domestic refuse into a rich harvest of fish and fresh vegetables. For example, the Mudially Fishermen's Cooperative Society is a collective of 300 families that lease 70 hectares into which wastewater from the city is released. Through a series of natural treatment processes – including the use of Eichhornia crassipes and other plants for absorbing oil, grease and heavy metals – the Cooperative has turned the area into a thriving fish farm and nature park. In 2005/06, the Cooperative sold fish worth over US$135,000 and shared income of more than US$55,000 among its members.

Reservoirs of Biodiversity

Wetland systems' rich biodiversity is becoming a focal point at International Treaty Conventions and within the World Wildlife Fund organization due to the high number of species present in wetlands, the small global geographic area of wetlands, the number of species which are endemic to wetlands, and the high productivity of wetland systems. Hundred of thousands of animal species, 20,000 of them vertebrates, are living in wetland systems. The discovery rate of fresh water fish is at 200 new species per year.

Biodiverse river basins: The Amazon holds 3,000 species of fresh water fish species within the boundaries of its basin whose function it is to disperse the seeds of trees. One of its key species, the Piramutaba catfish, *Brachyplatystoma vaillantii*, migrates more than 3,300 km (2,051 miles) from its nursery grounds near the mouth of the Amazon River to its spawning grounds in Andean tributaries (400 m or 437 yards above sea level) distributing plants seed along the route.

Productive intertidal zones: Intertidal mudflats have a similar productivity even while possessing a low number of species. The abundance of invertebrates found within the mud are a food source for migratory waterfowl.

Critical life-stage habitat: Mudflats, saltmarshes, mangroves, and seagrass beds contain bother species richness and productivity, and are home to important nursery areas for many commercial fish stocks.

Genetic Diversity: Many species in wetland systems are unique due to the long period of time that the ecosystem has been physically isolated from other aquatic sources. The number of endemic species in Lake Baikal in Russia classifies it as a hotspot for biodiversity and one of the most biodiverse wetlands in the entire world.

Lake Baikal: Evidence from a research study by Mazepova *et al.* suggest that the number of crustacean species endemic to Baikal Lake (>690 species and subspecies) exceeds the number of the same groups of animals inhabiting all the fresh water bodies of Eurasia together. Its 150 species of free-living Platyhelminthes alone is analogous to the entire number in all of Eastern Siberia. The 34 species and subspecies number of Baikal sculpins is more than twice the number of the analogous fauna that inhabits Eurasia. One of most exciting discoveries was made by A.V. Shoshin who registered about 300 species of free-living nematodes using only 6 near-shore sampling localities in the Southern Baikal. "If we will take into consideration, that about 60 % of the animals can be found nowhere else except Baikal, it may be assumed that the lake may be the biodiversity centre of the Eurasian continent."

Human Impact: Biodiversity loss occurs in wetland systems through land use changes, habitat destruction, pollution, exploitation of resources, and invasive species. Vulnerable, threatened, and endangered species number at 17% of waterfowl, 38% of fresh-water dependent mammals, 33% of fresh water fish, 26% of fresh water amphibians, 72% of fresh water turtles, 86% of marine turtles, 43% of crocodilians and 27% of coral reef-building species.

The impact of maintaining biodiversity is seen at the local level through job creation, sustainability, and community productivity. A good example is the Lower Mekong basin which runs through Cambodia, Laos, and Vietnam. Supporting over 55 million people, the sustainability of the region is enhanced through wildlife tours. The US state of Florida has estimated that US$ 1.6 billion was generated in state revenue from recreational activities associated with wildlife. Sustainable harvesting for medicinal remedies found in native wetlands plants in the Caribbean and Australia include the Red Mangrove *Rhizophora mangle* which possesses antibacterial, wound-healing, anti-ulcer effects, and antioxidant properties.

Wetland Products

Wetland systems naturally produce an array of vegetation and other ecological products that can harvested for personal and commercial use. The most significant of these is fish which have all or part of their life-cycle occur within a wetland system. Fresh and saltwater fish are the main source of protein for one billion people and comprise 15% of an additional two billion people's diets. In addition, fish generate a fishing industry that provides 80% of the income and employment to residents in developing countries. Another food staple found in wetland systems is rice, a popular grain that is consumed at the rate of 1/5 of the total global calorie count. In Bangladesh, Cambodia and Vietnam, where rice paddies are predominant on the landscape, rice consumption reach 70%.

Food converted to sweeteners and carbohydrates include the sago palm of Asia and Africa (cooking oil), the nipa palm of Asia (sugar, vinegar, alcohol, and fodder) and honey collection from mangroves. More than supplemental dietary intake, this produce sustains entire villages. Coastal Thailand villages earn the key portion of their income from sugar production while the country of Cuba relocates more than 30,000 hives each year to track the seasonal flowering of the mangrove *Avicennia*.

Other mangrove-derived products:

- Fuelwood
- Salt (produced by evaporating seawater)
- Animal fodder
- Traditional medicines (e.g. from mangrove bark)
- Fibres for textiles
- Dyes and tannins

Human Impact: Over-fishing is the major problem for sustainable use of wetlands. The field of aquaculture within the fisheries industries is eliminating mass areas of wetland systems through practices seen such as in the shrimp farming industry's destruction of mangroves. Aquaculture is continuing to develop rapidly throughout the Asia-Pacific region specifically in China with world holdings in Asia equal to 90% of the total number of aquaculture farms and 80% of its global value. Threats to rice fields mainly stem from inappropriate water management, introduction of invasive alien species, agricultural fertilizers, pesticides, and land use changes. Industrial-scale production of palm oil threatens the biodiversity of wetland ecosystems in parts of south-east Asia, Africa, and other developing countries. Exploitation can occur at the community level as is sometimes seen throughout coastal villages of Southern Thailand where each resident may obtain for themselves every consumable of the mangrove forest (fuelwood, timber, honey, resins, crab, and shellfish) which then becomes threatened through increasing population and continual harvest. Other issues that occur on a global level include an uneven contribution to climate change, point and non-point pollution, and air and water quality issues due to destructive wetland practices.

Wetlands and Climate Change

Wetlands perform two important functions in relation to climate change. They have mitigation effects through their ability to sink carbon, and adaptation effects through their ability to store and regulate water. Wetlands store approximately 44.6 Tg C y^{-1} globally. In salt marshes and mangrove swamps in particular, the average carbon sequestration rate is 201g CO_2 m^{-2} y^{-1} while peatlands sequester approximately 20-30g CO_2 m^{-2} y^{-1}.

Carbon sequestration has sponsored blue carbon initiatives. More on blue carbon and carbon sequestration.

They also are emitters of nitrous oxide, a major greenhouse gas.

Scientific Projections

"Low water and occasional drying of the wetland bottom during droughts (dry marsh phase) stimulate plant recruitment from a diverse seed bank and increase productivity by mobilizing nutrients. In contrast, high water during deluges (lake marsh phase) causes turnover in plant populations and creates greater interspersion of element cover and open water, but lowers overall productivity. During a cover cycle that ranges from open water to complete vegetation cover, annual net primary productivity may vary 20-fold."

Nitrous Oxide Production from Wetland Soils

Coastal wetlands, such as tropical mangroves and temperate salt marshes are known to be sinks of carbon, therefore mitigating climate change, however they are also emitters of nitrous oxide (N_2O), which is a greenhouse gas with a global warming potential 300 times that of carbon dioxide and the dominant ozone depleting substance emitted in the 21st century. Anthropogenic greenhouse gas (GHG) emissions have rapidly increased in the atmosphere due to the combustion of fossil fuels and deforestation practices, and these gases are major contributors to global climate change.

Although wetlands act as natural buffers towards nutrients expelled from surrounding watersheds, excess nutrients mainly through anthropogenic sources have been shown to significantly increase the N_2O fluxes from their soils through denitrification and nitrification processes. Anthropogenic sources of nutrients into waterways have increased substantially causing eutrophication especially in coastal systems. The main sources of coastal eutrophication are industrially made nitrogen, which is used as fertilizer in agricultural practices, as well as septic waste runoff. Nitrogen (N) is the limiting nutrient for photosynthetic processes in saline systems, however in excess, it can lead to an overproduction of organic matter that then leads to hypoxic and anoxic zones within the water column. Without oxygen, other organisms cannot survive, including economically important finfish and shellfish species. A study in the intertidal region of a New England salt marsh showed that excess levels of nutrients might increase N_2O emissions rather than sequester them.

Nitrous oxide fluxes from wetlands in the southern hemisphere is lacking as are ecosystem based studies including the role of dominant organisms that alter sediment biogeochemistry. Aquatic invertebrates produce ecologically relevant nitrous oxide emissions emissions due to ingestion of denitrifying bacteria that live within the subtidal sediment and water column and thus may also be influencing nitrous oxide production within wetlands.

Conservation

Wetlands have historically been the victim of large draining efforts for real estate development, or flooding for use as recreational lakes. Since the 1970s, more focus has been put on preserving wetlands for their natural function yet by 1993 half the world's wetlands had been drained. Wetlands provide a valuable flood control function. Wetlands are very effective at filtering and cleaning water pollution, (often from agricultural runoff from the farms that replaced the wetlands in the first place). To replace these wetland ecosystem services enormous amounts of money had to be spent on water purification plants, along with the remediation measures for controlling floods: dam and levee construction.

In order to produce sustainable wetlands, short-term, private-sector profits need to come secondary to global equity. Decision-makers must valuate wetland type, provided ecosystem service, long-term benefit, and current subsidies inflating valuation on either the private or public sector side. Analysis using the impact of hurricanes versus storm protection features projected wetland valuation at US$33,000/hectare/year.

Balancing Wetland Conservation with the Needs of People

Wetlands are vital ecosystems that provide livelihoods for the millions of people who live in and around them. The Millennium Development Goals (MDGs) called for different sectors to join forces to secure wetland environments in the context of sustainable development and improving human wellbeing. A three-year project carried out by Wetlands International in partnership with the International Water Management Institute found that it is possible to conserve wetlands while improving the livelihoods of people living among them. Case studies conducted in Malawi and Zambia looked at how dambos – wet, grassy valleys or depressions where water seeps to the surface – can be farmed sustainably to improve livelihoods. Mismanaged or overused dambos often become degraded, however, using a knowledge exchange between local farmers and environmental managers, a protocol was developed using soil and water management practices. Project outcomes included a high yield of crops, development of sustainable farming techniques, and adequate water management generating enough water for use as irrigation. Before the project, there were cases where people had died from starvation due to food shortages. By the end of it, many more people had access to enough water to grow vegetables. A key achievement was that villagers had secure food supplies during long, dry months. They also benefited in other ways: nutrition was improved by growing a wider range of crops, and villagers could also invest in health and education by selling produce and saving money.

Ramsar Convention

The Convention on Wetlands of International Importance, especially as Waterfowl Habitat, or Ramsar Convention, is an international treaty designed to address global concerns regarding wetland loss and degradation. The primary purposes of the treaty are to list wetlands of international importance and to promote their wise use, with the ultimate goal of preserving the world's wetlands. Methods include restricting access to the majority portion of wetland areas, as well as educating the public to combat the misconception that wetlands are wastelands. The Convention works closely with five International Organisation Partners.

These are: Birdlife International, IUCN, International Water Management Institute, Wetlands International and World Wide Fund for Nature. The partners provide technical expertise, help conduct or facilitate field studies and provide financial support. The IOPs also participate regularly as observers in all meetings of the Conference of the Parties and the Standing Committee and as full members of the Scientific and Technical Review Panel.

Valuation

The value of a wetland system to the earth and to humankind is one of the most important valuations that can be computed for sustainable development. A guideline involving assessing a wetland, keeping inventories of known wetlands, and monitoring the same wetlands over time is the current process that is used to educate environmental decision-makers such as governments on the importance of wetland protection and conservation.

Important Considerations

- Constructed Wetlands take 10–100 years to fully resemble the vegetative composition of a natural wetland.
- Artificial wetlands do not have hydric soil. The soil has very low levels of organic carbon and total nitrogen compared to natural wetland systems.
- Organic matter can be added to degraded natural wetlands to help restore their productivity before the wetland is destroyed.

Assessment

Rapid assessment: Five steps to assessing a wetland:

1. Collect general biodiversity data in order to inventory and prioritize wetland species, communities and ecosystems. Obtain baseline biodiversity information for a given area.
2. Gather information on the status of a focus or target species such as threatened species. Collect data pertaining to the conservation of a specific species.

3. Gain information on the effects of human or natural disturbance (changes) on a given area or species.
4. Gather information that is indicative of the general ecosystem health or condition of a specific wetland ecosystem.
5. Determine the potential for sustainable use of biological resources in a particular wetland ecosystem.

Inventory

Developing a global inventory of wetlands has proven to be a large and difficult undertaking. Current efforts are based on available data, but both classification and spatial resolution have proven to be inadequate for regional or site-specific environmental management decision-making. It is difficult to identify small, long, and narrow wetlands within the landscape. Many of today's remote sensing satellites do not have sufficient spatial and spectral resolution to monitor wetland conditions, although multispectral IKONOS and QuickBird data may offer improved spatial resolutions once it is 4 m or higher. Majority of the pixels are just mixtures of several plant species or vegetation types and are difficult to isolate which translates into an inability to classify the vegetation that defines the wetland. Improved remote sensing information, coupled with good knowledge domain on wetlands will facilitate expanded efforts in wetland monitoring and mapping. This will also be extremely important because we expect to see major shifts in species composition due to both anthropogenic land use and natural changes in the environment caused by climate change.

Monitoring

Mapping

A wetland system needs to be monitored over time to in order to assess whether it is functioning at an ecologically sustainable level or whether it is becoming degraded. Degraded wetlands will suffer a loss in water quality, a high number of threatened and endangered species, and poor soil conditions.

Due to the large size of wetlands, mapping is an effective tool to monitor wetlands. There are many remote sensing methods that can be used to map wetlands. Remote-sensing technology permits the acquisition of timely digital data on a repetitive basis. This repeat coverage allows wetlands, as well as the adjacent land-cover and land-use types, to be monitored seasonally and/or annually. Using digital data provides a standardized data-collection procedure and an opportunity for data integration within a geographic information system. Traditionally, Landsat 5 Thematic Mapper (TM), Landsat 7 Enhanced Thematic Mapper Plus (ETM +), and the SPOT 4 and 5 satellite systems have been used for this purpose. More recently, however, multispectral IKONOS and QuickBird data, with spatial resolutions of 4 m by 4 m and 2.44 m by 2.44 m, respectively, have been shown to be excellent sources of data when mapping and monitoring smaller wetland habitats and vegetation communities.

For example, Detroit Lakes Wetland Management District assessed area wetlands in Michigan, USA using remote sensing. Through using this technology, satellite images were taken over a large geographic area and extended period. In addition, using this technique was less costly and time-consuming compared to the older method using visual interpretation of aerial photographs. In comparison, most aerial photographs also require experienced interpreters to extract information based on structure and texture while the interpretation of remote sensing data only requires analysis of one characteristic (spectral).

However, there are a number of limitations associated with this type of image acquisition. Analysis of wetlands has proved difficult because to obtain the data it is often linked to other purposes such as the analysis of land cover or land use. Practically, many natural wetlands are difficult to monitor as these areas are quite often difficult to access and require exposure to native wildlife and potential endemic disease.

Further Improvements

Methods to develop a classification system for specific biota of interest could assist with technological advances that will allow for identification at a very high accuracy rate. The issue of the cost and expertise involved in remote sensing technology is still a factor hindering further advancements in image acquisition and data processing. Future improvements in current wetland vegetation mapping could include the use of more recent and better geospatial data when it is available.

List of Wetland Types

Wetland types:

- A—Marine and Coastal Zone wetlands
    1. Marine waters—permanent shallow waters less than six metres deep at low tide; includes sea bays, straits
    2. Subtidal aquatic beds; includes kelp beds, seagrasses, tropical marine meadows

3. Coral reefs
4. Rocky marine shores; includes rocky offshore islands, sea cliffs
5. Sand, shingle or pebble beaches; includes sand bars, spits, sandy islets
6. Intertidal mud, sand or salt flats
7. Intertidal marshes; includes saltmarshes, salt meadows, saltings, raised salt marshes, tidal brackish and freshwater marshes
8. Intertidal forested wetlands; includes mangrove swamps, nipa swamps, tidal freshwater swamp forests
9. Brackish to saline lagoons and marshes with one or more relatively narrow connections with the sea
10. Freshwater lagoons and marshes in the coastal zone
11. Non-tidal freshwater forested wetlands

- B—Inland wetlands

1. Permanent rivers and streams; includes waterfalls
2. Seasonal and irregular rivers and streams
3. Inland deltas (permanent)
4. Riverine floodplains; includes river flats, flooded river basins, seasonally flooded grassland, savanna and palm savanna
5. Permanent freshwater lakes (> 8 ha); includes large oxbow lakes
6. Seasonal/intermittent freshwater lakes (> 8 ha), floodplain lakes
7. Permanent saline/brackish lakes
8. Seasonal/intermittent saline lakes
9. Permanent freshwater ponds (< 8 ha), marshes and swamps on inorganic soils; with emergent vegetation waterlogged for at least most of the growing season
10. Seasonal/intermittent freshwater ponds and marshes on inorganic soils; includes sloughs, potholes; seasonally flooded meadows, sedge marshes
11. Permanent saline/brackish marshes
12. Seasonal saline marshes
13. Shrub swamps; shrub-dominated freshwater marsh, shrub carr, alder thicket on inorganic soils

14. Freshwater swamp forest; seasonally flooded forest, wooded swamps; on inorganic soils
15. Peatlands; forest, shrub or open bogs
16. Alpine and tundra wetlands; includes alpine meadows, tundra pools, temporary waters from snow melt
17. Freshwater springs, oases and rock pools
18. Geothermal wetlands
19. Inland, subterranean karst wetlands

- C—Human-made wetlands

1. Water storage areas; reservoirs, barrages, hydro-electric dams, impoundments (generally > 8 ha)
2. Ponds, including farm ponds, stock ponds, small tanks (generally < 8 ha)
3. Aquaculture ponds; fish ponds, shrimp ponds
4. Salt exploitation; salt pans, salines
5. Excavations; gravel pits, borrow pits, mining pools
6. Wastewater treatment; sewage farms, settling ponds, oxidation basins
7. Irrigated land and irrigation channels; rice fields, canals, ditches
8. Seasonally flooded arable land, farm land

Chapter 3

Ecology of Lakes, Pond and Stream

A lake is an area, (prototypically filled with water, also of variable size), localized in a basin, that is surrounded by land apart from any river or other outlet that serves to feed or drain the lake. Lakes lie on land and are not part of the ocean, and therefore are distinct from lagoons, and are also larger and deeper than ponds. Lakes can be contrasted with rivers or streams, which are usually flowing. However most lakes are fed and drained by rivers and streams.

Natural lakes are generally found in mountainous areas, rift zones, and areas with ongoing glaciation. Other lakes are found in endorheic basins or along the courses of mature rivers. In some parts of the world there are many lakes because of chaotic drainage patterns left over from the last Ice Age. All lakes are temporary over geologic time scales, as they will slowly fill in with sediments or spill out of the basin containing them.

Many lakes are artificial and are constructed for industrial or agricultural use, for hydro-electric power generation or domestic water supply, or for aesthetic or recreational purposes.

There is considerable uncertainty about defining the difference between lakes and ponds, and no current internationally accepted definition of either term across scientific disciplines or political boundaries exists. For example, limnologists have defined lakes as water bodies which are simply a larger version of a pond, which can have wave action on the shoreline or where wind-induced turbulence plays a major role in mixing the water column. None of these definitions completely excludes ponds and all are difficult to measure. For this reason there has been increasing use made of simple size-based definitions to separate ponds and lakes. One definition of *lake* is a body of water of 2 hectares (5 acres) or more in area; however, others have defined lakes as waterbodies of 5 hectares (12 acres) and above, or 8 hectares (20 acres) and above. Charles Elton, one of the founders of ecology, regarded lakes as waterbodies of 40 hectares (99 acres) or more. The term *lake* is also used to describe a feature such as Lake Eyre, which is a dry basin most of the time but may become filled under seasonal conditions of heavy rainfall.

In common usage many lakes bear names ending with the word *pond*, and a lesser number of names ending with *lake* are in quasi-technical fact, ponds. One textbook illustrates this point with the following: "In Newfoundland, for example, almost every lake is called a pond, whereas in Wisconsin, almost every pond is called a lake."

One hydrology book proposes to define it as a body of water with the following five chacteristics:

- it partially or totally fills one or several basins connected by straits
- has essentially the same water level in all parts (except for relatively short-lived variations caused by wind, varying ice cover, large inflows, etc.)
- it does not have regular intrusion of sea water
- a considerable portion of the sediment suspended in the water is captured by the basins (for this to happen they need to have a sufficiently small inflow-to-volume ratio)
- the area measured at the mean water level exceeds an arbitrarily chosen threshold (for instance, one hectare)

With the exception of the sea water intrusion criterion, the other ones have been accepted or elaborated upon by other hydrology publications.

Distribution of Lakes

The majority of lakes on Earth are fresh water, and most lie in the Northern Hemisphere at higher latitudes. Canada, with a deranged drainage system has an estimated 31,752 lakes larger than 3 square kilometres (1.2 sq mi) and an unknown total number of lakes, but is estimated to be at least 2 million. Finland has 187,888 lakes 500 square metres (5,400 sq ft) or larger, of which 56,000 are large (10,000 square metres (110,000 sq ft) or larger).

Most lakes have at least one natural outflow in the form of a river or stream, which maintain a lake's average level by allowing the drainage of excess water. Some lakes do not have a natural outflow and lose water solely by evaporation or underground seepage or both. They are termed endorheic lakes.

Many lakes are artificial and are constructed for hydro-electric power generation, aesthetic purposes, recreational purposes, industrial use, agricultural use or domestic water supply.

Evidence of extraterrestrial lakes exists; "definitive evidence of lakes filled with methane" was announced by NASA as returned by the Cassini Probe observing the moon Titan, which orbits the planet Saturn.

Globally, lakes are greatly outnumbered by ponds: of an estimated 304 million standing water bodies worldwide, 91% are 1 hectare (2.5 acres) or less in area. Small lakes are also much more numerous than large lakes: in terms of area, one-third of the world's standing water is represented by lakes and ponds of 10 hectares (25 acres) or less. However, large lakes account for much of the area of standing water with 122 large lakes of 1,000 square kilometres (390 sq mi, 100,000 ha, 247,000 acres) or more representing about 29% of the total global area of standing inland water.

Origin of Natural Lakes

There are a number of natural processes that can form lakes. A recent tectonic uplift of a mountain range can create bowl-shaped depressions that accumulate water and form lakes. The advance and retreat of glaciers can scrape depressions in the surface where water accumulates; such lakes are common in Scandinavia, Patagonia, Siberia and Canada. The most notable examples are probably the Great Lakes of North America.

Lakes can also form by means of landslides or by glacial blockages. An example of the latter occurred during the last ice age in the U.S. state of Washington, when a huge lake formed behind a glacial flow; when the ice retreated, the result was an immense flood that created the Dry Falls at Sun Lakes, Washington.

Salt lakes (also called saline lakes) can form where there is no natural outlet or where the water evaporates rapidly and the drainage surface of the water table has a higher-than-normal salt content. Examples of salt lakes include Great Salt Lake, the Aral Sea and the Dead Sea.

Small, crescent-shaped lakes called oxbow lakes can form in river valleys as a result of meandering. The slow-moving river forms a sinuous shape as the outer side of bends are eroded away more rapidly than the inner side. Eventually a horseshoe bend is formed and the river cuts through the narrow neck. This new passage then forms the main passage for the river and the ends of the bend become silted up, thus forming a bow-shaped lake.

Crater lakes are formed in volcanic craters and calderas which fill up with precipitation more rapidly than they empty via evaporation. Sometimes the latter are called caldera lakes, although often no distinction is made. An example is Crater Lake in Oregon, located within the caldera of Mount Mazama. The caldera was created in a massive volcanic eruption that led to the subsidence of Mount Mazama around 4860 BC.

Gloe Lakes are freshwater lakes that have emerged when the water they consists of has been separated, not considerably long before, from the sea as a consequence of post-glacial rebound.

Some lakes, such as Lake Jackson in Florida, USA, come into existence as a result of sinkhole activity.

Lake Vostok is a subglacial lake in Antarctica, possibly the largest in the world. The pressure from the ice atop it and its internal chemical composition mean that, if the lake were drilled into, a fissure could result that would spray somewhat like a geyser.

Most lakes are geologically young and shrinking since the natural results of erosion will tend to wear away the sides and fill the basin. Exceptions are those such as Lake Baikal and Lake Tanganyika that lie along continental rift zones and are created by the crust's subsidence as two plates are pulled apart. These lakes are the oldest and deepest in the world. Lake Baikal, which is 25-30 million years old, is deepening at a faster rate than it is being filled by erosion and may be destined over millions of years to become attached to the global ocean. The Red Sea, for example, is thought to have originated as a rift-valley lake.

Types of Lakes

- *Periglacial lake:* Part of the lake's margin is formed by an ice sheet, ice cap or glacier, the ice having obstructed the natural drainage of the land.
- *Subglacial lake:* A lake which is permanently covered by ice. They can occur under glaciers, ice caps or ice sheets. There are many such lakes, but Lake Vostok in Antarctica is by far the

largest. They are kept liquid because the overlying ice acts as a thermal insulator retaining energy introduced to its underside by friction, by water percolating through crevasses, by the pressure from the mass of the ice sheet above or by geothermal heating below.

- *Glacial lake:* a lake with origins in a melted glacier, such as a kettle lake.
- Artificial lake: A lake created by flooding land behind a dam, called an impoundment or reservoir, by deliberate human excavation, or by the flooding of an excavation incident to a mineral-extraction operation such as an open pit mine or quarry. Some of the world's largest lakes are reservoirs like Hirakud Dam in India.
- *Endorheic lake, terminal or closed:* A lake which has no significant outflow, either through rivers or underground diffusion. Any water within an endorheic basin leaves the system only through evaporation or seepage. These lakes, such as Lake Eyre in central Australia or the Aral Sea in central Asia, The Great Salt Lake in Western United States, are most common in desert locations.
- *Meromictic lake:* A lake which has layers of water which do not intermix. The deepest layer of water in such a lake does not contain any dissolved oxygen. The layers of sediment at the bottom of a meromictic lake remain relatively undisturbed because there are no living aerobic organisms.
- *Fjord lake:* A lake in a glacially eroded valley that has been eroded below sea level.
- *Oxbow lake:* A lake which is formed when a wide meander from a stream or a river is cut off to form a lake. They are called "oxbow" lakes due to the distinctive curved shape that results from this process.
- *Rift lake or sag pond:* A lake which forms as a result of subsidence along a geological fault in the Earth's tectonic plates. Examples include the Rift Valley lakes of eastern Africa and Lake Baikal in Siberia.
- *Underground lake:* A lake which is formed under the surface of the Earth's crust. Such a lake may be associated with caves, aquifers or springs.
- *Crater lake:* A lake which forms in a volcanic caldera or crater after the volcano has been inactive for some time. Water in this

type of lake may be fresh or highly acidic, and may contain various dissolved minerals. Some also have geothermal activity, especially if the volcano is merely dormant rather than extinct.

- *Lava lake:* A pool of molten lava contained in a volcanic crater or other depression. Lava lakes that have partly or completely solidified are also referred to as lava lakes.
- *Former:* A lake which is no longer in existence. Such lakes include prehistoric lakes and lakes which have permanently dried up through evaporation or human intervention. Owens Lake in California, USA, is an example of a former lake. Former lakes are a common feature of the Basin and Range area of southwestern North America.
- *Ephemeral lake, intermittent lake, or seasonal lake:* A seasonal lake that exists as a body of water during only part of the year.
- *Shrunken:* Closely related to *former* lakes, a shrunken lake is one which has drastically decreased in size over geological time. Lake Agassiz, which once covered much of central North America, is a good example of a shrunken lake. Two notable remnants of this lake are Lake Winnipeg and Lake Winnipegosis.
- *Eolic lake:* A lake which forms in a depression created by the activity of the winds.
- Vlei, in South Africa, shallow lakes which vary considerably with seasons.
- *Epishelf lakes*, unique lakes which exist on top of a dense saltwater body and are surrounded by ice. These are mostly found in the Antarctica.

Characteristics

Lakes have numerous features in addition to lake type, such as drainage basin (also known as catchment area), inflow and outflow, nutrient content, dissolved oxygen, pollutants, pH, and sedimentation.

Changes in the level of a lake are controlled by the difference between the input and output compared to the total volume of the lake. Significant input sources are precipitation onto the lake, runoff carried by streams and channels from the lake's catchment area, groundwater channels and aquifers, and artificial sources from outside the catchment area. Output sources are evaporation from the lake, surface and groundwater flows, and any extraction of lake water by humans. As climate conditions and human water requirements vary, these will create fluctuations in the lake level.

Lakes can be also categorized on the basis of their richness in nutrients, which typically affect plant growth. Nutrient-poor lakes are said to be *oligotrophic* and are generally clear, having a low concentration of plant life. *Mesotrophic lakes* have good clarity and an average level of nutrients. *Eutrophic* lakes are enriched with nutrients, resulting in good plant growth and possible algal blooms. *Hypertrophic* lakes are bodies of water that have been excessively enriched with nutrients. These lakes typically have poor clarity and are subject to devastating algal blooms. Lakes typically reach this condition due to human activities, such as heavy use of fertilizers in the lake catchment area. Such lakes are of little use to humans and have a poor ecosystem due to decreased dissolved oxygen.

Due to the unusual relationship between water's temperature and its density, lakes form layers called thermoclines, layers of drastically varying temperature relative to depth. Fresh water is most dense at about 4 degrees Celsius (39.2 °F) at sea level. When the temperature of the water at the surface of a lake reaches the same temperature as deeper water, as it does during the cooler months in temperate climates, the water in the lake can mix, bringing oxygen-starved water up from the depths and bringing oxygen down to decomposing sediments. Deep temperate lakes can maintain a reservoir of cold water year-round, which allows some cities to tap that reservoir for deep lake water cooling.

Since the surface water of deep tropical lakes never reaches the temperature of maximum density, there is no process that makes the water mix. The deeper layer becomes oxygen starved and can become saturated with carbon dioxide, or other gases such as sulphur dioxide if there is even a trace of volcanic activity. Exceptional events, such as earthquakes or landslides, can cause mixing which rapidly brings the deep layers up to the surface and release a vast cloud of gas which lay trapped in solution in the colder water at the bottom of the lake. This is called a limnic eruption. An example is the disaster at Lake Nyos in Cameroon. The amount of gas that can be dissolved in water is directly related to pressure. As deep water surfaces, the pressure drops and a vast amount of gas comes out of solution. Under these circumstances carbon dioxide is hazardous because it is heavier than air and displaces it, so it may flow down a river valley to human settlements and cause mass asphyxiation.

The material at the bottom of a lake, or *lake bed*, may be composed of a wide variety of inorganics, such as silt or sand, and organic material, such as decaying plant or animal matter. The composition of the lake bed has a significant impact on the flora and fauna found within the lake's environs by contributing to the amounts and the types of nutrients available.

A paired (black and white) layer of the varved lake sediments correspond to a year. During winter, when organisms die, carbon is deposited down, resulting to a black layer. At the same year, during summer, only few organic materials are deposited, resulting to a white layer at the lake bed. These are commonly used to track past paleontological events.

Natural lakes provide a microcosm of living and nonliving elements that are relatively independent of their surrounding environments. Therefore, lake organisms can often be studied in isolation from the lake's surroundings.

Lake Ecosystem

The ecosystem of a lake includes biotic (living) plants, animals and micro-organisms, as well as abiotic (nonliving) physical and chemical interactions. Lake ecosystems are a prime examples of lentic ecosystems. *Lentic* refers to standing or relatively still water, from the Latin *lentus*, which means sluggish. Lentic waters range from ponds to lakes to wetlands, and much of this article applies to lentic ecosystems in general. Lentic ecosystems can be compared with lotic ecosystems, which involve flowing terrestrial waters such as rivers and streams. Together, these two fields form the more general study area of freshwater or aquatic ecology.

Lentic systems are diverse, ranging from a small, temporary rainwater pool a few inches deep to Lake Baikal, which has a maximum depth of 1740 m. The general distinction between pools/ponds and lakes is vague, but Brown states that ponds and pools have their entire bottom surfaces exposed to light, while lakes do not. In addition, some lakes become seasonally stratified (discussed in more detail below.) Ponds and pools have two regions: the pelagic open water zone, and the benthic zone, which comprises the bottom and shore regions. Since lakes have deep bottom regions not exposed to light, these systems have an additional zone, the profundal. These three areas can have very different abiotic conditions and, hence, host species that are specifically adapted to live there.

Important Abiotic Factors of Lakes

Light: Light provides the solar energy required to drive the process of photosynthesis, the major energy source of lentic systems. The amount of light received depends upon a combination of several factors. Small ponds may experience shading by surrounding trees, while cloud cover may affect light availability in all systems, regardless of size. Seasonal and diurnal considerations also play a role in light availability because the shallower the angle at which light strikes water, the more light is lost by reflection.

This is known as Beer's law. Once light has penetrated the surface, it may also be scattered by particles suspended in the water column. This scattering decreases the total amount of light as depth increases. Lakes are divided into photic and aphotic regions, the prior receiving sunlight and latter being below the depths of light penetration, making it void of photosynthetic capacity. In relation to lake zonation, the pelagic and benthic zones are considered to lie within the photic region, while the profundal zone is in the aphotic region.

Temperature: Temperature is an important abiotic factor in lentic ecosystems because most of the biota are poikilothermic, where internal body temperatures are defined by the surrounding system. Water can be heated or cooled through radiation at the surface and conduction to or from the air and surrounding substrate. Shallow ponds often have a continuous temperature gradient from warmer waters at the surface to cooler waters at the bottom. In addition, temperature fluctuations can be very great in these systems, both diurnally and seasonally.

Temperature regimes are very different in large lakes. In temperate regions, for example, as air temperatures increase, the icy layer formed on the surface of the lake breaks up, leaving the water at approximately 4 °C. This is the temperature at which water has the highest density. As the season progresses, the warmer air temperatures heat the surface waters, making them less dense.

The deeper waters remain cool and dense due to reduced light penetration. As the summer begins, two distinct layers become established, with such a large temperature difference between them that they remain stratified. The lowest zone in the lake is the coldest and is called the hypolimnion. The upper warm zone is called the epilimnion. Between these zones is a band of rapid temperature change called the thermocline. During the colder fall season, heat is lost at the surface and the epilimnion cools. When the temperatures of the two zones are close enough, the waters begin to mix again to create a uniform temperature, an event termed lake turnover. In the winter, inverse stratification occurs as water near the surface cools freezes, while warmer, but denser water remains near the bottom. A thermocline is established, and the cycle repeats.

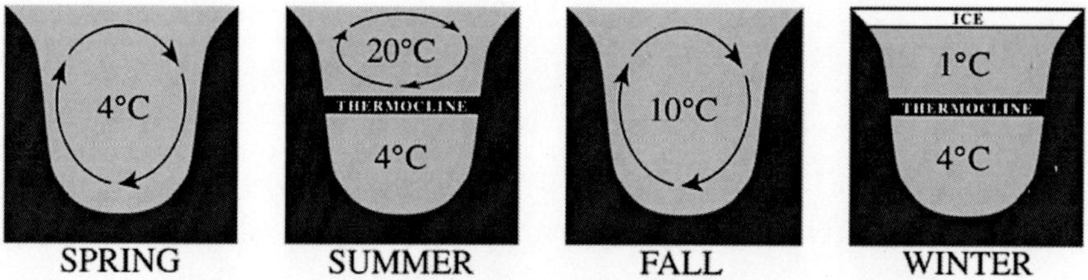

Figure: *Seasonal stratification in temperate lakes*

Wind

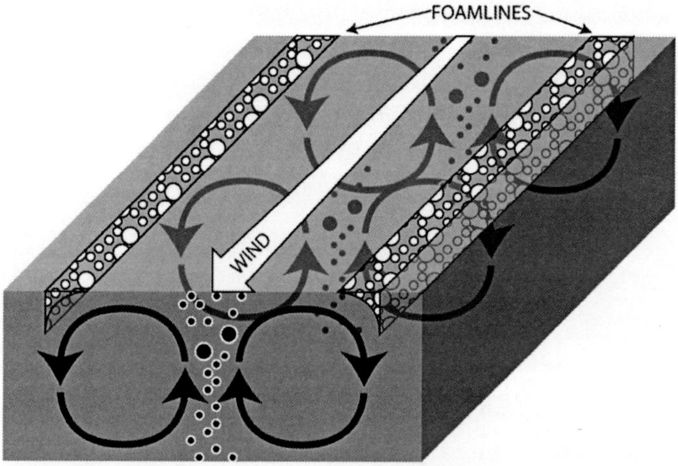

Figure: *Illustration of Langmuir rotations; open circles=positively buoyant particles, closed circles=negatively buoyant particles*

In exposed systems, wind can create turbulent, spiral-formed surface currents called Langmuir circulations. Exactly how these currents become established is still not well understood, but it is evident that it involves some interaction between horizontal surface currents and surface gravity waves. The visible result of these rotations, which can be seen in any lake, are the surface foamlines that run parallel to the wind direction.

Positively buoyant particles and small organisms concentrate in the foamline at the surface and negatively buoyant objects are found in the upwelling current between the two rotations. Objects with neutral buoyancy tend to be evenly distributed in the water column. This turbulence circulates nutrients in the water column, making it crucial for many pelagic species, however its effect on benthic and profundal organisms is minimal to non-existent, respectively. The degree of nutrient circulation is system specific, as it depends upon such factors as wind strength and duration, as well as lake or pool depth and productivity.

Chemistry

Oxygen is essential for organismal respiration. The amount of oxygen present in standing waters depends upon:

1) the area of transparent water exposed to the air,
2) the circulation of water within the system and
3) the amount of oxygen generated and used by organisms present. In shallow, plant-rich pools there may be great fluctuations of oxygen, with extremely high concentrations occurring during the day due to photosynthesis and very low values at night when respiration is the dominant process of primary producers. Thermal stratification in larger systems can also affect the amount of oxygen present in different zones. The epilimnion is oxygen rich because it circulates quickly, gaining oxygen via contact with the air. The hypolimnion, however, circulates very slowly and has no atmospheric contact. Additionally, fewer green plants exist in the hypolimnion, so there is less oxygen released from photosynthesis. In spring and fall when the epilimnion and hypolimnion mix, oxygen becomes more evenly distributed in the system. Low oxygen levels are characteristic of the profundal zone due to the accumulation of decaying vegetation and animal matter that "rains" down from the pelagic and benthic zones and the inability to support primary producers.

Phosphorus is important for all organisms because it is a component of DNA and RNA and is involved in cell metabolism as a component of ATP and ADP. Also, phosphorus is not found in large quantities in freshwater systems, limiting photosynthesis in primary producers, making it the main determinant of lentic system production. The phosphorus cycle is complex, but the model outlined below describes the basic pathways. Phosphorus mainly enters a pond or lake through runoff from the watershed or by atmospheric deposition. Upon entering the system, a reactive form of phosphorus is usually taken up by algae and macrophytes, which release a non-reactive phosphorus compound as a byproduct of photosynthesis. This phosphorus can drift downwards and become part of the benthic or profundal sediment, or it can be remineralized to the reactive form by microbes in the water column. Similarly, non-reactive phosphorus in the sediment can be remineralized into the reactive form. Sediments are generally richer in phosphorus than lake water, however, indicating that this nutrient may have a long residency time there before it is remineralized and re-introduced to the system.

Aquatic Respiration

Aquatic respiration is the process whereby an aquatic animal obtains oxygen from water.

Respiratory Systems

Fish: In most fish respiration takes place through gills. Lungfish, however, possess one or two lungs. The labyrinth fish have developed a special organ that allows them to take advantage of the oxygen of the air, but is not a true lung. Fish use the process known as countercurrent flow, in which water and blood flow in opposite directions across the gills, maximizing the diffusion of oxygen.

Mollusks: Mollusks generally possess gills that allow exchange of oxygen from an aqueous environment into the circulatory system. These animals also possess a heart that pumps blood which contains hemocyaninine as its oxygen-capturing molecule. Therefore, this respiratory system is similar to that of vertebrate fish. The respiratory system of gastropods can include either gills or a lung.

Arthropods: Aquatic arthropods generally possess some form of gills in which gas exchange takes place by diffusing through the exoskeleton. Others may breathe atmospheric air while remaining submerged, via breathing tubes or trapped air bubbles, though some aquatic insects may remain submerged indefinitely and respire using a plastron.

Aquatic reptiles: The anatomical structure of the lungs is less complex in reptiles than in mammals, with reptiles lacking the very extensive airway tree structure found in mammalian lungs. Gas exchange in reptiles still occurs in alveoli however, reptiles do not possess a diaphragm. Thus, breathing occurs via a change in the volume of the body cavity which is controlled by contraction of intercostal muscles in all reptiles except turtles. In turtles, contraction of specific pairs of flank muscles governs inspiration or expiration.

Amphibians: Both the lungs and the skin serve as respiratory organs in amphibians. The skin of these animals is highly vascularized and moist, with moisture maintained via secretion of mucus from specialized cells. While the lungs are of primary importance to breathing control, the skin's unique properties aid rapid gas exchange when amphibians are submerged in oxygen-rich water.

Aquatic birds: The respiratory system of birds differs significantly from that found in mammals, containing unique anatomical features such as air sacs. The lungs of birds also do not have the capacity to inflate as birds lack a diaphragm and a pleural cavity. Gas exchange in birds occurs between air capillaries and blood capillaries, rather than in alveoli.

Gills: Large aquatic animals have developed gills for respiration which are specifically adapted to their function, for example, they have:

- A large surface area to allow as much oxygen to enter the gills as possible due to the fact that more of the gas comes into contact with the membrane
- Good blood supply to maintain the concentration gradient needed
- Thin membrane to allow for a short diffusion pathway
- each gill arch has two rows (hemibranchs) of gill filaments
- each gill filament has many lamellae

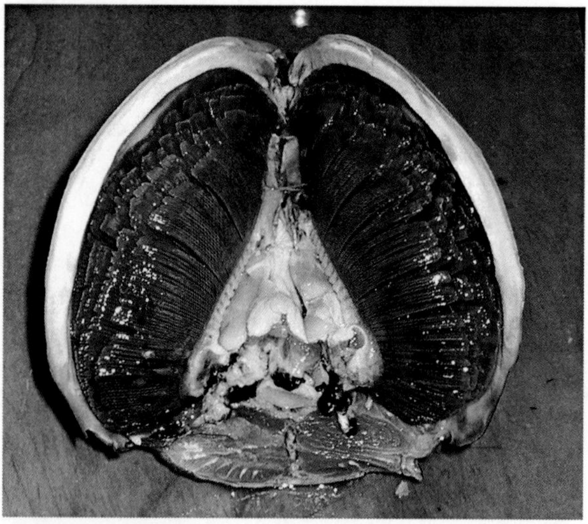

Figure: *Posterior view of the gills of a tuna*

In osteichthyes, the gills contain 4 gill arches on each side of the head, two on each side for chondrichthyes or 7 gill baskets on each side of the fish's head in Lampreys In fish, the long bony cover for the gill (the operculum) can be used for pushing water. Some fish pump water using the operculum. Without an operculum, other methods, such as ventilation, are required. Some species of sharks use this system. When they swim, water flows into the mouth and across the gills. Because these sharks rely on this technique, they must keep swimming in order to respire.

Bony fish use countercurrent flow to maximize the intake of oxygen that can diffuse through the gill. Countercurrent flow occurs when deoxygenated blood moves through the gill in one direction while oxygenated water moves through the gill in the opposite direction. This mechanism maintains the concentration gradient thus increasing the efficiency of the respiration process as well and prevents the oxygen levels from reaching an equilibrium. Cartilaginous fish do not have a countercurrent flow system as they lack bones which are needed to have the opened out gill that bony fish have.

Control of Respiration

Scientists have investigated what part of the body is responsible for maintaining the respiratory rhythm. They found that neurons located in the brainstem of fish are responsible for the genesis of the respiratory rhythm.

The position of these neurons is slightly different from the centres of respiratory genesis in mammals but they are located in the same brain compartment, which has caused debates about the homology of respiratory centres between aquatic and terrestrial species. In both aquatic and terrestrial respiration, the exact mechanisms by which neurons can generate this involuntary rhythm are still not completely understood.

Another important feature of the respiratory rhythm is that it is modulated to adapt to the oxygen consumption of the body. As observed in mammals, fish "breathe" faster and heavier when they do physical exercise. The mechanisms by which these changes occur have been strongly debated over more than 100 years between scientists. The authors can be classified in 2 schools:

1. Those who think that the major part of the respiratory changes are pre-programmed in the brain, which would imply that neurons from locomotion centres of the brain connect to respiratory centres in anticipation of movements.
2. Those who think that the major part of the respiratory changes result from the detection of muscle contraction, and that respiration is adapted as a consequence of muscular contraction and oxygen consumption.

   This would imply that the brain possesses some kind of detection mechanisms that would trigger a respiratory response when muscular contraction occurs.

Many now agree that both mechanisms are probably present and complementary, or working alongside a mechanism that can detect changes in oxygen and/or carbon dioxide blood saturation.

Lentic System Biota

Bacteria

Bacteria are present in all regions of lentic waters. Free-living forms are associated with decomposing organic material, biofilm on the surfaces of rocks and plants, suspended in the water column, and in the sediments of the benthic and profundal zones. Other forms are also associated with the guts of lentic animals as parasites or in commensal relationships. Bacteria play an important role in system metabolism through nutrient recycling, which is discussed in the Trophic Relationships section.

Primary Producers

Algae, including both phytoplankton and periphyton are the principle photosynthesizers in ponds and lakes. Phytoplankton are found drifting in the water column of the pelagic zone. Many species have a higher density than water which should make them sink and end up in the benthos. To combat this, phytoplankton have developed density changing mechanisms, by forming vacuoles and gas vesicles or by changing their shapes to induce drag, slowing their descent. A very sophisticated adaptation utilized by a small number of species is a tail-like flagellum that can adjust vertical position and allow movement in any direction. Phytoplankton can also maintain their presence in the water column by being circulated in Langmuir rotations. Periphytic algae, on the other hand, are attached to a substrate. In lakes and ponds, they can cover all benthic surfaces. Both types of plankton are important as food sources and as oxygen providers.

Aquatic plants live in both the benthic and pelagic zones and can be grouped according to their manner of growth: 1) emergent = rooted in the substrate but with leaves and flowers extending into the air, 2) floating-leaved = rooted in the substrate but with floating leaves, 3) submersed = growing beneath the surface and 4) free-floating macrophytes = not rooted in the substrate and floating on the surface. These various forms of macrophytes generally occur in different areas of the benthic zone, with emergent vegetation nearest the shoreline, then floating-leaved macrophytes, followed by submersed vegetation. Free-floating macrophytes can occur anywhere on the system's surface.

Aquatic plants are more buoyant than their terrestrial counterparts because freshwater has a higher density than air. This makes structural rigidity unimportant in lakes and ponds (except in the aerial stems and leaves).

Thus, the leaves and stems of most aquatic plants use less energy to construct and maintain woody tissue, investing that energy into fast growth instead. In order to contend with stresses induced by wind and waves, plants must be both flexible and tough. Light, water depth and substrate types are the most important factors controlling the distribution of submerged aquatic plants. Macrophytes are sources of food, oxygen, and habitat structure in the benthic zone, but cannot penetrate the depths of the euphotic zone and hence are not found there.

Invertebrates

Figure: *Water striders are predatory insects which rely on surface tension to walk on top of water. They live on the surface of ponds, marshes, and other quiet waters. They can move very quickly, up to 1.5 m/s.*

Zooplankton are tiny animals suspended in the water column. Like phytoplankton, these species have developed mechanisms that keep them from sinking to deeper waters, including drag-inducing body forms and the active flicking of appendages such as antennae or spines. Remaining in the water column may have its advantages in terms of feeding, but this zone's lack of refugia leaves zooplankton vulnerable to predation. In response, some species, especially Daphnia sp., make daily vertical migrations in the water column by passively sinking to the darker lower depths during the day and actively moving towards the surface during the night. Also, because conditions in a lentic system can be quite variable across seasons, zooplankton have the ability to switch from laying regular eggs to resting eggs when there is a lack of food, temperatures fall below 2 °C, or if predator abundance is high.

These resting eggs have a diapause, or dormancy period that should allow the zooplankton to encounter conditions that are more favourable to survival when they finally hatch. The invertebrates that inhabit the benthic zone are numerically dominated by small species and are species rich compared to the zooplankton of the open water. They include Crustaceans (e.g. crabs, crayfish, and shrimp), molluscs (e.g. clams and snails), and numerous types of insects. These organisms are mostly found in the areas of macrophyte growth, where the richest resources, highly oxygenated water, and warmest portion of the ecosystem are found. The structurally diverse macrophyte beds are important sites for the accumulation of organic matter, and provide an ideal area for colonization. The sediments and plants also offer a great deal of protection from predatory fishes.

Very few invertebrates are able to inhabit the cold, dark, and oxygen poor profundal zone. Those that can are often red in colour due to the presence of large amounts of hemoglobin, which greatly increases the amount of oxygen carried to cells. Because the concentration of oxygen within this zone is low, most species construct tunnels or borrows in which they can hide and make the minimum movements necessary to circulate water through, drawing oxygen to them without expending much energy.

Fish and Other Vertebrates

Fish have a range of physiological tolerances that are dependent upon which species they belong to. They have different lethal temperatures, dissolved oxygen requirements, and spawning needs that are based on their activity levels and behaviours. Because fishes are highly mobile, they are able to deal with unsuitable abiotic factors in one zone by simply moving to another. A detrital feeder in the profundal zone, for example, that finds the oxygen concentration has dropped too low may feed closer to the benthic zone. A fish might also alter its residence during different parts of its life history: hatching in a sediment nest, then moving to the weedy benthic zone to develop in a protected environment with food resources, and finally into the pelagic zone as an adult.

Other vertebrate taxa inhabit lentic systems as well. These include amphibians (e.g. salamanders and frogs), reptiles (e.g. snakes, turtles, and alligators), and a large number of waterfowl species. Most of these vertebrates spend part of their time in terrestrial habitats and thus are not directly affected by abiotic factors in the lake or pond. Many fish species are important as consumers and as prey species to the larger vertebrates mentioned above.

Trophic Relationships

Primary Producers

Lentic systems gain most of their energy from photosynthesis performed by aquatic plants and algae. This autochthonous process involves the combination of carbon dioxide, water, and solar energy to produce carbohydrates and dissolved oxygen. Within a lake or pond, the potential rate of photosynthesis generally decreases with depth due to light attenuation. Photosynthesis, however, is often low at the top few millimetres of the surface, likely due to inhibition by ultraviolet light. The exact depth and photosynthetic rate measurements of this curve are system specific and depend upon:

1) the total biomass of photosynthesizing cells,
2) the amount of light attenuating materials and
3) the abundance and frequency range of light absorbing pigments (i.e. chlorophylls) inside of photosynthesizing cells.

The energy created by these primary producers is important for the community because it is transferred to higher trophic levels via consumption.

Bacteria

The vast majority of bacteria in lakes and ponds obtain their energy by decomposing vegetation and animal matter. In the pelagic zone, dead fish and the occasional allochthonous input of litterfall are examples of coarse particulate organic matter (CPOM>1 mm). Bacteria degrade these into fine particulate organic matter (FPOM<1 mm) and then further into usable nutrients. Small organisms such as plankton are also characterized as FPOM. Very low concentrations of nutrients are released during decomposition because the bacteria are utilizing them to build their own biomass. Bacteria, however, are consumed by protozoa, which are in turn consumed by zooplankton, and then further up the trophic levels. Nutrients, including those that contain carbon and phosphorus, are reintroduced into the water column at any number of points along this food chain via excretion or organism death, making them available again for bacteria. This regeneration cycle is known as the microbial loop and is a key component of lentic food webs.

The decomposition of organic materials can continue in the benthic and profundal zones if the matter falls through the water column before being completely digested by the pelagic bacteria. Bacteria are found in the greatest abundance here in sediments, where they are typically 2-1000 times more prevalent than in the water column.

Benthic Invertebrates

Benthic invertebrates, due to their high level of species richness, have many methods of prey capture. Filter feeders create currents via siphons or beating cilia, to pull water and its nutritional contents, towards themselves for straining. Grazers use scraping, rasping, and shredding adaptations to feed on periphytic algae and macrophytes. Members of the collector guild browse the sediments, picking out specific particles with raptorial appendages. Deposit feeding invertebrates indiscriminately consume sediment, digesting any organic material it contains. Finally, some invertebrates belong to the predator guild, capturing and consuming living animals. The profundal zone is home to a unique group of filter feeders that use small body movements to draw a current through burrows that they have created in the sediment. This mode of feeding requires the least amount of motion, allowing these species to conserve energy. A small number of invertebrate taxa are predators in the profundal zone. These species are likely from other regions and only come to these depths to feed. The vast majority of invertebrates in this zone are deposit feeders, getting their energy from the surrounding sediments.

Fish

Fish size, mobility, and sensory capabilities allow them to exploit a broad prey base, covering multiple zonation regions. Like invertebrates, fish feeding habits can be categorized into guilds. In the pelagic zone, herbivores graze on periphyton and macrophytes or pick phytoplankton out of the water column. Carnivores include fishes that feed on zooplankton in the water column (zooplanktivores), insects at the water's surface, on benthic structures, or in the sediment (insectivores), and those that feed on other fish (piscivores). Fish that consume detritus and gain energy by processing its organic material are called detritivores. Omnivores ingest a wide variety of prey, encompassing floral, faunal, and detrital material. Finally, members of the parasitic guild acquire nutrition from a host species, usually another fish or large vertebrate. Fish taxa are flexible in their feeding roles, varying their diets with environmental conditions and prey availability. Many species also undergo a diet shift as they develop. Therefore, it is likely that any single fish occupies multiple feeding guilds within its lifetime.

Lentic Food Webs

As noted in the previous sections, the lentic biota are linked in complex web of trophic relationships. These organisms can be considered to loosely be associated with specific trophic groups (e.g. primary producers, herbivores, primary carnivores, secondary carnivores, etc.). Scientists have developed several theories in order to understand the mechanisms that control the abundance and diversity within these groups. Very generally, top-down processes dictate that the abundance of prey taxa is dependent upon the actions of consumers from higher trophic levels. Typically, these processes operate only between two trophic levels, with no effect on the others. In some cases, however, aquatic systems experience a trophic cascade; for example, this might occur if primary producers experience less grazing by herbivores because these herbivores are suppressed by carnivores. Bottom-up processes are functioning when the abundance or diversity of members of higher trophic levels is dependent upon the availability or quality of resources from lower levels. Finally, a combined regulating theory, bottom-up:top-down, combines the predicted influences of consumers and resource availability. It predicts that trophic levels close to the lowest trophic levels will be most influenced by bottom-up forces, while top-down effects should be strongest at top levels.

Community patterns and diversity

Local Species Richness

The biodiversity of a lentic system increases with the surface area of the lake or pond. This is attributable to the higher likelihood of partly terrestrial species of finding a larger system. Also, because larger systems typically have larger populations, the chance of extinction is decreased. Additional factors, including temperature regime, pH, nutrient availability, habitat complexity, speciation rates, competition, and predation, have been linked to the number of species present within systems.

Succession Patterns in Plankton Communities – The PEG Model

Phytoplankton and zooplankton communities in lake systems undergo seasonal succession in relation to nutrient availability, predation, and competition. Sommer *et al.* described these patterns as part of the Plankton Ecology Group (PEG) model, with 24 statements constructed from the analysis of numerous systems. The following includes a subset of these statements, as explained by Brönmark and Hansson illustrating succession through a single seasonal cycle:

Winter:

1. Increased nutrient and light availability result in rapid phytoplankton growth towards the end of winter. The dominant

species, such as diatoms, are small and have quick growth capabilities.

2. These plankton are consumed by zooplankton, which become the dominant plankton taxa.

Spring

3. A clear water phase occurs, as phytoplankton populations become depleted due to increased predation by growing numbers of zooplankton.

Summer

4. Zooplankton abundance declines as a result of decreased phytoplankton prey and increased predation by juvenile fishes.
5. With increased nutrient availability and decreased predation from zooplankton, a diverse phytoplankton community develops.
6. As the summer continues, nutrients become depleted in a predictable order: phosphorus, silica, and then nitrogen. The abundance of various phytoplankton species varies in relation to their biological need for these nutrients.
7. Small-sized zooplankton become the dominant type of zooplankton because they are less vulnerable to fish predation.

Fall

8. Predation by fishes is reduced due to lower temperatures and zooplankton of all sizes increase in number.

Winter

9. Cold temperatures and decreased light availability result in lower rates of primary production and decreased phytoplankton populations. 10. Reproduction in zooplankton decreases due to lower temperatures and less prey.

The PEG model presents an idealized version of this succession pattern, while natural systems are known for their variation.

Latitudinal Patterns

There is a well-documented global pattern that correlates decreasing plant and animal diversity with increasing latitude, that is to say, there are fewer species as one moves towards the poles. The cause of this pattern is one of the greatest puzzles for ecologists today. Theories for its explanation include energy availability, climatic variability, disturbance, competition, etc. Despite this global diversity gradient, this pattern can be weak for freshwater systems compared to global marine and terrestrial systems. This may be related to size, as Hillebrand and Azovsky found that smaller organisms (protozoa and plankton) did not follow the expected trend strongly, while larger species (vertebrates) did. They attributed this to better dispersal ability by smaller organisms, which may result in high distributions globally.

Natural Lake Lifecycles

Lake Creation

Lakes can be formed in a variety of ways, but the most common are discussed briefly below. The oldest and largest systems are the result of tectonic activities. The rift lakes in Africa, for example are the result of seismic activity along the site of separation of two tectonic plates. Ice-formed lakes are created when glaciers recede, leaving behind abnormalities in the landscape shape that are then filled with water. Finally, oxbow lakes are fluvial in origin, resulting when a meandering river bend is pinched off from the main channel.

Natural Extinction

All lakes and ponds receive sediment inputs. Since these systems are not really expanding, it is logical to assume that they will become increasingly shallower in depth, eventually becoming wetlands or terrestrial vegetation. The length of this process should depend upon a combination of depth and sedimentation rate. Moss gives the example of Lake Tanganyika, which reaches a depth of 1500 m and has a sedimentation rate of 0.5 mm/yr. Assuming that sedimentation is not influenced by anthropogenic factors, this system should go extinct in approximately 3 million years. Shallow lentic systems might also fill in as swamps encroach inward from the edges. These processes operate on a much shorter timescale, taking hundreds to thousands of years to complete the extinction process.

Human Impacts

Acidification

Sulphur dioxide and nitrogen oxides are naturally released from volcanoes, organic compounds in the soil, wetlands, and marine systems, but the majority of these compounds come from the combustion of coal, oil, gasoline, and the smelting of ores containing sulphur. These substances dissolve in atmospheric moisture and enter lentic systems as acid rain. Lakes and ponds that contain bedrock that is rich in carbonates have a natural buffer, resulting in no alteration of pH. Systems without this bedrock, however, are very sensitive to acid inputs because they have a low neutralizing capacity, resulting in pH declines even with only small inputs of acid. At a pH of 5–6 algal species diversity and biomass decrease considerably, leading to an increase in water transparency – a characteristic feature of acidified lakes. As the pH continues lower, all fauna becomes less diverse. The most significant feature is the disruption of fish reproduction.

Thus, the population is eventually composed of few, old individuals that eventually die and leave the systems without fishes. Acid rain has been especially harmful to lakes in Scandinavia, western Scotland, west Wales and the north eastern United States.

Eutrophication

Eutrophic systems contain a high concentration of phosphorus (~30+μg/L), nitrogen (~1500+μg/L), or both. Phosphorus enters lentic waters from wastewater treatment effluents, discharge from raw sewage, or from runoff of farmland. Nitrogen mostly comes from agricultural fertilizers from runoff or leaching and subsequent groundwater flow. This increase in nutrients required for primary producers results in a massive increase of phytoplankton growth, termed a plankton bloom.

This bloom decreases water transparency, leading to the loss of submerged plants. The resultant reduction in habitat structure has negative impacts on the species' that utilize it for spawning, maturation and general survival. Additionally, the large number of short-lived phytoplankton result in a massive amount of dead biomass settling into the sediment. Bacteria need large amounts of oxygen to decompose this material, reducing the oxygen concentration of the water. This is especially pronounced in stratified lakes when the thermocline prevents oxygen rich water from the surface to mix with lower levels. Low or anoxic conditions preclude the existence of many taxa that are not physiologically tolerant of these conditions.

Invasive Species

Invasive species have been introduced to lentic systems through both purposeful events (e.g. stocking game and food species) as well as unintentional events (e.g. in ballast water). These organisms can affect natives via competition for prey or habitat, predation, habitat alteration, hybridization, or the introduction of harmful diseases and parasites. With regard to native species, invaders may cause changes in size and age structure, distribution, density, population growth, and may even drive populations to extinction. Examples of prominent invaders of lentic systems include the zebra mussel and sea lamprey in the Great Lakes.

How Lakes Disappear

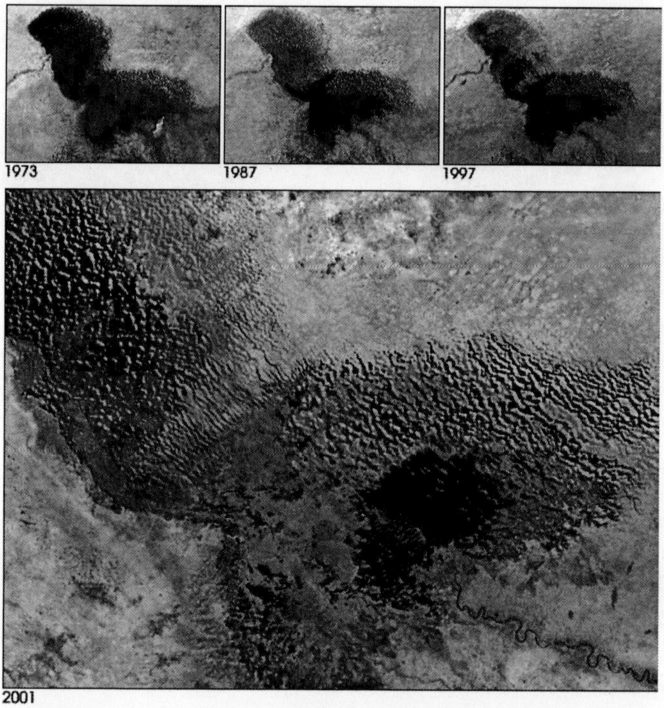

Figure: *Lake Chad in a 2001 satellite image, with the actual lake in blue, and vegetation on top of the old lake bed in green.*

Figure: *Lake Badwater, February 9, 2005. Landsat 5 satellite photo.*

The lake may be infilled with deposited sediment and gradually become a wetland such as a swamp or marsh. Large water plants, typically totos, accelerate this closing process significantly because they partially decompose to form peat soils that fill the shallows.

Figure: *Badwater Basin dry lake, February 15, 2007. Landsat 5 satellite photo.*

Conversely, peat soils in a marsh can naturally burn and reverse this process to recreate a shallow lake resulting in a dynamic equilibrium between marsh and lake. This is significant since wildfire has been largely suppressed in the developed world over the past century. This has artificially converted many shallow lakes into emergent marshes. Turbid lakes and lakes with many plant-eating fish tend to disappear more slowly.

A "disappearing" lake (barely noticeable on a human timescale) typically has extensive plant mats at the water's edge. These become a new habitat for other plants, like peat moss when conditions are right, and animals, many of which are very rare. Gradually the lake closes and young peat may form, forming a fen. In lowland river valleys where a river can meander, the presence of peat is explained by the infilling of historical oxbow lakes. In the very last stages of succession, trees can grow in, eventually turning the wetland into a forest.

Some lakes can disappear seasonally. These are called intermittent lakes, ephemeral lakes, or seasonal lakes and can be found in karstic terrain. A prime example of an intermittent lake is Lake Cerknica in Slovenia or Lag Prau Pulte in Graubünden. Other intermittent lakes are only the result of above-average precipitation in a closed, or endorheic basin, usually filling dry lake beds.

This can occur in some of the driest places on earth, like Death Valley. This occurred in the spring of 2005, after unusually heavy rains. The lake did not last into the summer, and was quickly evaporated. A more commonly filled lake of this type is Sevier Lake of west-central Utah.

Sometimes a lake will disappear quickly. On 3 June 2005, in Nizhny Novgorod Oblast, Russia, a lake called Lake Beloye vanished in a matter of minutes. News sources reported that government officials theorized that this strange phenomenon may have been caused by a shift in the soil underneath the lake that allowed its water to drain through channels leading to the Oka River.

The presence of ground permafrost is important to the persistence of some lakes. According to research published in the journal *Science* ("Disappearing Arctic Lakes", June 2005), thawing permafrost may explain the shrinking or disappearance of hundreds of large Arctic lakes across western Siberia. The idea here is that rising air and soil temperatures thaw permafrost, allowing the lakes to drain away into the ground.

Some lakes disappear because of human development factors. The shrinking Aral Sea is described as being "murdered" by the diversion for irrigation of the rivers feeding it.

Extraterrestrial Lakes

Figure: *Titan's north polar hydrocarbon seas and lakes as seen in a false-colour Cassini synthetic aperture radar mosaic.*

Only one world other than Earth is known to harbor large lakes, Saturn's largest moon, Titan. Photographs and spectroscopic analysis by the Cassini–Huygens spacecraft show liquid ethane on the surface, which is thought to be mixed with liquid methane. The largest Titanean lake, Kraken Mare at 400,000 km^2, is three-times the size of any lake on Earth, and even the second, Ligeia Mare, is estimated to be slightly larger than Earth's Lake Michigan–Huron.

Jupiter's large moon Io is volcanically active, and as a result sulphur deposits have accumulated on the surface. Some photographs taken during the Galileo mission appear to show lakes of liquid sulphur in volcanic caldera, though these are more analogous to lake of lava than of water on Earth.

The planet Mars is too cold and has too little atmospheric pressure to permit the pooling of liquid water. Geologic evidence appears to confirm, however, that ancient lakes once formed on the surface.

It is also possible that volcanic activity on Mars will occasionally melt subsurface ice, creating large temporary lakes. This water would quickly freeze and then sublimate, unless insulated in some manner, such as by a coating of volcanic ash.

There are dark basaltic plains on the Moon, similar to lunar maria but smaller, that are called *lacus* (singular *lacus*, Latin for "lake") because they were thought by early astronomers to be lakes of water.

Notable Lakes on Earth

- The deepest lake is Lake Baikal in Siberia, with a bottom at 1,637 metres (5,371 ft). Its mean depth is also the greatest in the world (749 metres (2,457 ft)).

  It is also the world's largest lake by volume (23,600 cubic kilometres (5,700 cu mi), though smaller than the Caspian Sea at 78,200 cubic kilometres (18,800 cu mi)), and the second longest (about 630 kilometres (390 mi) from tip to tip).
- The longest lake is Lake Tanganyika, with a length of about 660 kilometres (410 mi) (measured along the lake's centre line).

  It is also the second largest by volume and second deepest (1,470 metres (4,820 ft)) in the world, after lake Baikal.
- The world's oldest lake is Lake Baikal, followed by Lake Tanganyika (Tanzania).
- The world's highest lake, if size is not a criterion, may be the crater lake of Ojos del Salado, at 6,390 metres (20,965 ft).
- The highest large (greater than 250 square kilometres (97 sq mi)) lake in the world is the 290 square kilometres (110 sq mi) Pumoyong Tso (Pumuoyong Tso), in the Tibet Autonomous Region of China, at 28-34N 90-24E, 5,018 metres (16,463 ft) above sea level.
- The world's highest commercially navigable lake is Lake Titicaca in Peru and Bolivia at 3,812 m (12,507 ft). It is also the largest lake in South America.

- The world's lowest lake is the Dead Sea, bordering Israel and Jordan at 418 metres (1,371 ft) below sea level. It is also one of the lakes with highest salt concentration.
- Lake Michigan–Huron has the longest lake coastline in the world: about 5,250 kilometres (3,260 mi), excluding the coastline of its many inner islands. Even if it is considered two lakes, Lake Huron alone would still have the longest coastline in the world at 2,980 kilometres (1,850 mi).
- The largest island in a lake is Manitoulin Island in Lake Huron, with a surface area of 2,766 square kilometres (1,068 sq mi). Lake Manitou, located on Manitoulin Island, is the largest lake on an island in a lake.
- The largest lake located on an island is Nettilling Lake on Baffin Island, with an area of 5,542 square kilometres (2,140 sq mi) and a maximum length of 123 kilometres (76 mi).
- The largest lake in the world that drains naturally in two directions is Wollaston Lake.
- Lake Toba on the island of Sumatra is located in what is probably the largest resurgent caldera on Earth.
- The largest lake located completely within the boundaries of a single city is Lake Wanapitei in the city of Sudbury, Ontario, Canada. Before the current city boundaries came into effect in 2001, this status was held by Lake Ramsey, also in Sudbury.
- Lake Enriquillo in Dominican Republic is the only saltwater lake in the world inhabited by crocodiles.
- Lake Bernard, Ontario, Canada, claims to be the largest lake in the world with no islands.
- The largest lake in one country is Lake Michigan, in the U.S.A. However, it is sometimes considered part of Lake Michigan-Huron, making the record go to Great Bear Lake, Northwest Territories, in Canada, the largest lake within one jurisdiction.
- The largest lake on an island in a lake on an island is Crater Lake on Vulcano Island in Lake Taal on the island of Luzon, The Philippines.
- The northernmost named lake on Earth is Upper Dumbell Lake in the Qikiqtaaluk Region of Nunavut, Canada at a latitude of 82°28'N. It is located only 5.2 kilometres (3.2 mi) southwest of Alert, the northernmost settlement in the world. There are also several small lakes located north of Upper Dumbell Lake, but they are all unnamed and only appear on very detailed maps.

Largest by Continent

The largest lakes (surface area) by continent are:

- Australia – Lake Eyre (salt lake)
- Africa – Lake Victoria, also the third-largest freshwater lake on Earth. It is one of the Great Lakes of Africa.
- Antarctica – Lake Vostok (subglacial)
- Asia – Lake Baikal (if the Caspian Sea is considered a lake, it is the largest in Eurasia, but is divided between the two geographic continents)
- Oceania – Lake Eyre when filled; the largest permanent (and freshwater) lake in Oceania is Lake Taupo.
- Europe – Lake Ladoga, followed by Lake Onega, both located in northwestern Russia.
- North America – Lake Michigan-Huron, which is hydrologically a single lake. However, lakes Huron and Michigan are usually considered separate lakes, in which case Lake Superior would be the largest.
- South America – Lake Titicaca, which is also the highest navigable body of water on Earth at 3,812 metres (12,507 ft) above sea level. The much larger Lake Maracaibo is considered by some to be the second-oldest lake on Earth, but since it lies at sea level and nowadays is a contiguous body of water with the sea, others consider that it has turned into a small bay.

Pond

A pond is a body of standing water, either natural or man-made, that is usually smaller than a lake. They may arise naturally in floodplains as part of a river system, or they may be somewhat isolated depressions (examples include vernal pools and prairie potholes). Usually they contain shallow water with marsh and aquatic plants and animals. A few animals also make their home in ponds, including both alligators and beavers. The type of life in a pond is generally determined by a combination of factors including water level regime (particularly depth and duration of flooding) and nutrient levels, but other factors may also be important, including presence or absence of shading by trees, presence or absence of streams, effects of grazing animals, and salinity.

Humans also make ponds. A wide variety of man-made bodies of water are classified as ponds. Some ponds are created specifically for habitat restoration, including water treatment. Others, like water gardens, water features and koi ponds are designed for aesthetic ornamentation as landscape or architectural features. Fish ponds are designed for commercial fish breeding, and solar ponds designed to store thermal energy.

Standing bodies of water such as puddles, ponds, and lakes are often categorized separately from flowing water courses, such as a brook, creek, stream or river. Nutrient levels and water quality in natural or man-made ponds can be controlled through natural process such as algal growth, or man-made filtration such as an algae scrubber.

Technical Definitions

The technical distinction between a pond and a lake has not been universally standardized. Limnologists and freshwater biologists have proposed formal definitions for *pond*, in part to include 'bodies of water where light penetrates to the bottom of the waterbody,' 'bodies of water shallow enough for rooted water plants to grow throughout,' and 'bodies of water which lack wave action on the shoreline.' Each of these definitions has met with resistance or disapproval, as the defining characteristics are each difficult to measure or verify. Accordingly, some organizations and researchers have settled on technical definitions of *pond* and *lake* which rely on size alone.

Even among organizations and researchers who distinguish lakes from ponds by size alone, there is no universally recognised standard for the maximum size of a pond. The international Ramsar wetland convention sets the upper limit for pond size as 8 hectares (20 acres), but biologists have not universally adopted this convention. Researchers for the British charity Pond Conservation have defined a *pond* to be 'a man-made or natural waterbody which is between 1 m^2 and 20,000 m^2 in area (2 ha or ~5 acres), which holds water for four months of the year or more.' Other European biologists have set the upper size limit at 5 ha (12 acres).

In practice, a body of water is called a pond or a lake on an individual basis, as conventions change from place to place and over time. In North America, even larger bodies of water have been called ponds; for example, Walden Pond in Concord, Massachusetts measures 61 acres (25 ha), nearby Spot Pond is 340 acres (140 ha), while in between is Crystal Lake at 33 acres (13 ha). There are numerous examples in other states of bodies of water less than 10 acres (4.0 ha) being called lakes. As the case with Crystal Lake shows, marketing purposes may be the driving factor behind some names.

Formation

Ponds can result from a wide range of natural processes. Any depression in the ground which collects and retains a sufficient amount of precipitation can be considered a pond, and such depressions can be formed by a variety of geological and ecological events. Rivers often leave behind ponds in natural flood plains after spring flooding, and these can be very important to breeding fish, particularly in large river systems like the Amazon.

Retreating glaciers can leave behind landscapes filled with small depressions, each developing its own pond; an example is the Prairie Pothole Region of North America. Many areas of landscape contain small depressions which form temporary ponds after spring snow melt, or during rainy seasons; these are called vernal ponds, and may be important sites for amphibian breeding. Some ponds are created by animals. Beaver ponds are the best known example, but alligators also excavate ponds as well. In landscapes with organic soils, fires can also create depressions during periods of drought; these become open water when normal water levels return.

Characteristics

One of the most important features of ponds is the presence of standing water, which provides habitat for wetland plants and animals. Familiar examples might include water-lilies, frogs, turtles and herons. Often, the entire margin of the pond is fringed by wetland, and these wetlands support the aquatic food web, provide shelter for wildlife, and stabilize the shore of the pond. Some grazing animals like geese and muskrats consume the wetland plants directly as a source of food. In many other cases, however, the pond plants fall into the water and decay. A large number of invertebrates then feed on the decaying plants, and these invertebrates provide food for wetland species including fish, dragonflies and herons. The open water may allow algae to grow, and these algae may support yet another food web that includes aquatic insects and minnows. A pond, therefore, may have combinations of three different food webs, one based on larger plants, one based upon decayed plants, and one based upon algae. Hence, ponds often have a large number of different animal species using the wide array of food sources. They therefore provide an important source of biological diversity in landscapes.

Vernal ponds are ponds which dry up for part of the year. Naturally occurring vernal ponds do not usually have fish. They are called vernal ponds because they are typically at their peak depth in the spring ("vernal" means to do with the spring). The absence of fish is a very important characteristic, since it provides amphibians with breeding locations free from predation by fish. Hence, introducing fish to a pond can have serious detrimental consequences. In some parts of the world, such as California, the vernal ponds have rare and endangered plant species. On the coastal plain, they provide habitat for endangered frogs such as the Mississippi Gopher Frog.

Conservation and Management

Ponds, being small, are easily disrupted by human activity. Drainage of ponds is a frequent problem in agricultural areas, such as in the prairie potholes of North America. Although ponds are a useful source of water for cattle, overgrazing and wading can turn a pond into a muddy hole. Nutrient sources such as fertilized pastures, human sewage, and even lawn fertilizer can cause explosive growth of algae, and the loss of rooted plants and many other aquatic species. Roads near ponds can kill large numbers of amphibians and turtles that may migrate to and from the pond as part of their annual breeding cycle. Many well-intentioned people introduce fish to ponds, being unaware that some species of fish eat aquatic plants, stir up sediment and eat the young of amphibians and many invertebrate species. The gentle slope of land into ponds also provides an expanse of habitat for wetland plants and wet meadows. The construction of retaining walls, or lawns, can severely degrade the life in a pond.

In some landscapes, ponds are artificially constructed, perhaps to provide wildlife viewing opportunities, or to treat wastewater, or as part of a golf course. The design of a pond determines how productive it will be for wildlife. In general, gently sloping shorelines with broad expanses of wetland plants not only provide the best conditions for wildlife, but they help protect water quality from sources in the surrounding landscapes. It is also beneficial to allow water levels to fall each year during drier periods. Roads and houses should be kept as far away as possible.

Another important way to add ponds back into landscapes is to restore rivers so that they can flood and meander to create large numbers of natural ponds, including vernal pools and wetlands, in river valleys.

Nomenclature

In origin, pond is a variant form of the word pound, meaning a confining enclosure. As straying cattle are enclosed in a pound so water is enclosed in a pond. In earlier times, ponds were man-made and utilitarian; as stew ponds, mill ponds and so on. The significance of this feature seems, in some cases, to have been lost when the word was carried abroad with emigrants. In the United States, natural pools are often called ponds. Ponds for a specific purpose keep the adjective, such as "stock pond", used for watering livestock.

Pond usually implies a quite small body of water, generally smaller than one would require a boat to cross. Another definition is that a pond is a body of water where even its deepest areas are reached by sunlight or where a human can walk across the entire body of water without being submerged. In some dialects of English, pond normally refers to small artificially created bodies of water.

Some regions of the United States define a pond as a body of water with a surface area of less than 10 acres (4.0 ha). Minnesota, known as the 'land of 10,000 lakes' is commonly said to distinguish lakes from ponds, bogs and other water features by this definition, but also says that a lake is distinguished primarily by wave action reaching the shore.

The term is also used for temporary accumulation of water from surface runoff (*ponded* water).

There are various regional names for naturally occurring ponds. In Scotland, one of the terms is lochan, which may also apply to a large body of water such as a lake. In North American prairies, they may be termed playas.

Uses

Globally, the most important service provided by a pond, at many scales, is the production of fish and other wildlife. These are often also a source of food for humans, as well as an important source of recreation. At the same time, these ponds help maintain water quality by recycling nutrients.

In the Indian subcontinent, Hindu temples usually have a pond nearby so that pilgrims can take baths. These ponds are considered sacred.

In medieval times in Europe, it was typical for many monastery and castles (small, partly self-sufficient communities) to have fish ponds. These are still common in Europe and in East Asia (notably Japan), where koi may be kept.

In agriculture, treatment ponds may reduce nutrients released downstream from the pond. They may also provide irrigation reservoirs at times of drought.

Pond Ecology

Pond ecology is best described as the interaction of the life in your pond with the environment that exists there. A shallow, nutrient rich pond, exposed to sunlight with little water flowing through it will be teeming with algae and aquatic plants. It may have very little animal life present because of low oxygen levels. In contrast a newly created, deep, spring fed pond may have little life of any kind in it because of low temperatures and lack of food supply.

All ponds age. A pond begins with mostly water, few nutrients, and little aquatic life. Over time the pond accumulates nutrients. This enrichment process is called eutrophication. The addition of nutrients stimulates the growth of aquatic life. These organisms live and grow and die. Their remains decay in the pond and the nutrients it took to grow them are released back into the water of the pond to keep the cycle going. Eventually, though, there will be an accumulation of material that resists decay and the pond will fill up. It will become a bog and someday will resemble dry land. The process of return to dry land can happen in a decade or may take centuries. As a pond owner your job is to slow the process down as much as possible. Some of the principles you can employ are described below.

Exclude Nutrients

Four basic elements are required to make aquatic organisms. They are carbon, oxygen, nitrogen, phosphorous. Of course it takes more than these to make even the simplest organism, but these are the materials required in abundance. To prevent the rapid aging of a pond (eutrophication) aim to exclude the rapid introduction of these, especially nitrogen and phosphorous. There are three practices that are particularly helpful in slowing the aging process.

Animal access to ponds or streams that feed ponds should be restricted to limit inputs of nitrogen and phosphorous from their wastes.

Buffers

Maintaining vegetation in all areas through which water must flow to reach the pond is very beneficial to the pond. Such buffers both slow water down and filter it. Slow moving water allows sediment to drop out of the water. Much phosphorous is attached to soil particles so that sedimentation is effective in keeping phosphorous out of the pond. Keeping sediment out of the pond also prevents the pond from being made shallow by filling it with sediment. This contributes directly to our primary objective keeping the pond from returning to dry land. A deeper pond, will also be a cooler pond. A general principle of biology is that lower temperatures slow the growth of organisms. So again the buffer area contributes to conditions that help slow the aging process for the pond.

Sedimentation

Another method of keeping sediment out of ponds is to provide a shallow pool at the inlet of the pond. Water passing through this pool on its way to the pond will have an opportunity to drop its sediment load in the pool. This pool should be of such dimensions that it can be easily cleaned with a backhoe from the shore of the pool. A sedimentation pool helps the pond in the same way that sediment removal by buffer strips does.

Limit Fertilization

When it is possible to decrease the use of fertilizer on turf or crops grown in the watershed area of the pond, the pond will receive a benefit. One of the reasons for this is that plants are never 100 percent efficient in their use of fertilizer elements. So even applying fertilizers at appropriate rates results in some elements, particularly nitrogen, being not used and moving off site. Reductions in fertilizer rates will decrease the amounts getting off site.

Maintain Ecological Balance

Ponds are most satisfactory when there is a complete and balanced food web in place. Starting at the top this means that planktonic algae are present in sufficient quantity to feed some zooplankton. The zooplankton in turn provides food for the smallest fish and aquatic insects. These in turn become prey for larger fish which finally may be taken by raccoons, bears, or fishermen.

Another part of ecological balance involves the higher plant community. Too many plants are discouraging to the pond owner and are also detrimental to the food side of the ecology just described. From the pond owners point of view, a pond full of vegetation presents a poor appearance, interferes with fishing, swimming, and boating. From the view point of aquatic life there are problems too.

Some aquatic plants are valuable in providing shade, hiding places for small fish, habitat for some aquatic insects and animals, as well as being a food source for some fish and animals. When the vegetation becomes excessive not only does the angler's hook get entangled, the bait is hidden from that trophy bass. The vegetation that hides the bait also hides his prey making the hunt unsuccessful. Such a pond decreases its capacity to produce fish.

Two other examples of excessive vegetation will serve to illustrate undesirable consequences than can occur. A pond completely covered with water lilies or lotus will so shade the pond that no other vegetation will grow under the water. Neither will their be light enough to grow planktonic algae. This will be a very unproductive pond for anything besides lilies. The other example is excessive growth of duck weed or Watermeal. When the whole pond surface is covered with these plants, again the light is shut off and the pond will contain little life beneath the surface. These plants also virtually eliminate oxygenation of the water by maintaining a complete separation of the water surface from the atmosphere. As a result such a pond becomes oxygen deficient to such an extent that any fish present are killed from lack of oxygen.

A seldom discussed problem in the management of aquatic vegetation is the potential to eliminate too much vegetation or to eliminate beneficial plants along with the targeted weeds. This is something to remember when considering weed control in ponds. Some helps are to treat the pond in parts over time, to use mechanical methods, or perhaps to use an appropriate number of grass carp to keep things "pruned up" instead of wiped out.

Maintain Water Flow

A discussion that occurs when a new pond is planned concerns the water supply for the pond. A pond with a continuous supply of water is almost always going to be a more satisfactory pond than one with an intermittent water supply. Ponds lose significant water by evaporation during the summer. Ponds with sufficient inflow stay full while the water level in others declines exposing an unattractive muddy beach around the perimeter of the pond.

The nutrient conditions in a pond with a continuous overflow are likely to be better, because excess nutrients will leave with the overflow water. In contrast the pond having intermittent flows only has a chance to purge excess nutrients during storm events. Such ponds are prone to accumulate nutrients much more rapidly than their overflowing cousins. The accumulation of nutrients leads to excessive vegetative growth of all kinds as was noted above.

Encourage Aeration

Oxygen in pond water is very beneficial to the overall health of the pond. The value to fish is obvious. Less obvious, but of great importance, is the ability of the pond to get rid of waste. The waste that occurs in the pond includes "deposits" from its animal life - fish and geese, waste material that enter with storm water runoff, as well as plant and animals that die in the pond. Aerobic bacteria work about 20 times faster than anaerobic bacteria in breaking this waste down and putting it into solution. Once in solution it can be flushed out or is available to grow new life.

Oxygenation of pond is quite interesting. It happens in two major ways. Plants and algae do photosynthesis during the day and wind adds oxygen at night. The oxygen plants produce is released into the pond water and maintains high levels of oxygen in the pond. That's why conditions that prevent light from entering the pond have to be monitored or disaster can occur such as the complete cover by Watermeal mentioned earlier. All the oxygen manufactured by the Watermeal is released to the atmosphere rather than into the pond water. Any part of the pond that is too dark for photosynthesis to occur is also likely to be oxygen deficient unless the pond is being mixed from top to bottom. A Secchi disk can be lowered into the water to check for visibility. The depth of disappearance is noted. The surface water above this disappearance depth will be oxygenated by photosynthesis while water below that depth must be mixed to receive oxygen.

The other method of getting oxygen into pond water occurs by oxygen exchange with the atmosphere at the surface of the pond. The rougher the surface the more rapid the exchange. Also the more deficient the oxygen content of the water the faster the exchange occurs. This process is important at night and is critical for the pond with a heavy load of plants and animals. At night the plants do respiration instead of photosynthesis the same as the animals. By dawn the pond may be oxygen deficient if atmospheric aeration is impeded by lack of wind or especially by a covered surface.

Pond Fishes

Although a large number of fish species grow successfully in ponds, only a restricted number of species are usually cultivated on commercial scale. Reasons for this restricted choice is obvious. Commercial pond culture basically aims at achieving maximum possible rate of fish production and profit through optimum utilization of the natural food and the supplementary feed which drastically limits the choice of fish species for pond cultivation. Some of the basic criteria for selection are discussed below.

Criteria for selection of suitable fish species:

- Adaptability to undrainable pond environment
- Faster growth rate

- Efficient utilizers of natural food resources of the pond
- Efficient converter of artificial feed
- Hardy and not easily susceptible to disease
- Easy to breed and rear the seed
- Prolonged breeding period or multiple breeding frequency
- Non-predaceous, planktophagous and preferably herbivorous and detritus feeder
- Compatability with other cultivable species of fish
- Palatable with high nutritive value
- High market demand and high price.

To find all these qualities in one fish species would be very unlikely. Therefore, the species having maximum required traits are considered to be desirable for cultivation in undrainable ponds. Carps fit well to these criteria and hence the most widely cultivated food fishes in South Asia are the quick-growing, non-predatory carps.

Fish species suitable for culture in undrainable ponds.There are two major systems of carp culture in Asia: the Chinese polyculture system where Chinese carps are cultured together, and the Indian composite fish culture system where the Indian major carps and Chinese carps are combined. In China, Chinese carps such as silver carp (Hypophthalmichthys molitrix), grass carp (Ctenopharyngodon idella), bighead carp (Aristichthys nobilis), mud carp (Cirrhinus molitorella), black carp (Mylopharyngodon piceus), and common carp (Cyprinus carpio) are cultured under polyculture system in ponds. Under the Indian system of composite fish culture in undrainable ponds, three Indian carps, viz. catla (Catla catla), rohu (Labeo rohita) and mrigal (Cirrhinus mrigala), two Chinese carps, viz. silver carp (Hypophthalmichthys molitrix) and grass carp (Ctenopharyngodon idella), and common carp (Cyprinus carpio) are extensively cultivated.

Catla

Catla is the fastest growing Indian major carp species and widely distributed throughout India, Nepal, Pakistan, Burma and Bangladesh. It inhabits the surface layer of water and feeds upon plankton. Adult stages are predominantly zooplankton feeder, occasionally taking in decaying macrovegetation, phytoplankton and smaller molluscs. It attains maturity in the second year of life and carry over 70 000 eggs per kg body weight. It naturally breeds in rivers during monsoon season and under control conditions in bundhs as well. It does not breed in ponds. However, it responds well to hypophysation techniques. Seeds are easily reared in undrainable ponds of relatively smaller size. Under composite fish culture in ponds it usually grows to over 1 kg in one year.

Rohu

Rohu is the natural inhabitant of river systems of India, Nepal, Pakistan, Bangladesh and Burma. In recent years it has been transplanted to many countries of the world including Sri Lanka, Mauritius, USSR, Japan, Philippines, Laos, Malaysia and Thailand. Normally it occupies the column region of the aquatic ecosystem and feeds mostly on vegetable matter including higher plants, detritus, etc. Like catla it naturally breeds in rivers and under special conditions in bundhs. Except by hypophysation to which it responds quickly, it never breeds in ponds. It attains sexual maturity during the second year. However, certain percentages of pond-reared specimens mature within one year. Fecundity varies from 226 000 to about 2 800 000 depending upon the size. Rohu spawns during the monsoon (April—September). Seeds collected from rivers or produced by bundh breeding or induced breeding are reared with ease in seasonal or perennial undrainable ponds. Under pond culture conditions it grows upto 900 g within one year.

Mrigal

Mrigal inhabits all the major river systems of India, Pakistan, Bangladesh and Burma. The adult fish feeds upon filamentous green algae, diatoms, pieces of higher plants, decayed vegetable, mud and detritus. It is basically a bottom feeder and hence suitable for cultivation with column and surface feeder carps in ponds. Mrigal usually attains maturity within 1 or 2 years depending upon the agroclimatic conditions of the location. Fecundity ranges between 124 000 to over 1 900 000 depending upon size. Spawning season is linked with the onset and duration of the southwest monsoon. It does not breed in ponds, but can be easily bred in bundhs and by hypophysation. It is now being induced to breed twice within the same spawning season. Rearing of seed is usually undertaken in seasonal or perennial undrainable ponds. Under pond culture conditions it grows to over 1 kg in one year.

Figure: *Catla (Catla catla)*

Figure: *Rohu (Labeo rohita)*

Figure: *Mrigal (Cirrhinus mriqala)*

Silver Carp

Silver carp is basically inhabitant of major river systems of South and Central China and in the Amur Basin of USSR from where it has been transplanted throughout the Indo-Pacific region including India. It is a surface dweller feeding mainly upon zooplankton during its early stages and gradually becomes predominantly a phytoplankton feeder. Its relatively longer branchiospines provide a fine filter capable of retaining planktonic organisms. It readily accepts supplementary feed like oil cakes and rice bran mixture in pond culture systems. It does not breed in pond condition.

Figure: *Silver Carp (Hypophthalmichthys molitrix)*

However, through the technique of hypophysation they are induced to breed in ponds during the monsoon season. Fecundity varies greatly with the size and agroclimatic condition. A fecundity range of 145 000 to 2 044 000 has been found from silver carp. It takes about 2–6 years to mature in China, whereas in India it matures very early, within 2 years. Males mature earlier than the females. In composite fish culture ponds it usually attains over 1.5 kg within one year of rearing. Seed rearing is done in smaller seasonal or perennial undrainable ponds with a high rate of survival.

Grass Carp

Figure: *Grass carp (Ctenopharyngodon idella)*

Grass carp is a native of the river systems of South-Central and North China, and the Amur river of USSR. Its suitability in aquaculture and biological control of aquatic weed infestation has resulted in wide-scale transplantation throughout the world. In early life it feeds on planktonic organisms and gradually switches over to macrophytes. They are voracious eaters and show distinct preference for vegetable food materials such as grass, leaves, weeds, etc. However, they also accept supplementary artificial feed materials.

Usually only a portion of ingested food is digested and the rest is voided in semidigested or undigested form which, in turn, becomes choice food for the bottom dweller common carp (Alikunhi, Sukumaran and Parameswaran, 1963). In China it takes about 3–4 years to achieve maturity whereas in India it usually takes 2 years. The total number of eggs range between 308 800 and 618 100 from the fishes weighing between 4.7 kg to 7.0 kg. The fish does not breed under pond condition and hence seed production is achieved through hypophysation. Growth mainly depends on the rate of feeding. Under optimum feeding rate it can grow over 5 kg in one year (Sinha and Gupta, 1975). Usually it grows to over 1.5 kg in composite fish culture ponds.

Common Carp

Originally a native of temperate region of Asia, especially China, the common carp is now the most domesticated and cultivated carp species throughout the world. It is an omnivorous bottom dweller subsisting mainly on benthic fauna and decaying vegetable matter. It frequently burrows the pond bottom in search of food. This habit of digging the pond bottom helps in maintaining the productivity of undrainable ponds and hence culture of common carp with other carp species is of great advantage. Moreover, it also feeds directly on the undigested excreta of grass carp. Growth mainly depends upon the bottom biota, stocking density and the rate of supplementary feed. In composite fish culture ponds it grows to about 1 kg within one year. In a tropical climate it spawns throughout the year in the pond environment with two peak periods, one from January to March and the other during July and August. Eggs are small and adhesive in nature. In tropical conditions it attains maturity within 12 months (Alikunhi, 1966).

Figure: *Common carp (Cyprinus carpio)*

Farm Ponds for Water, Fish and Livelihood

Smallholder farmers make up 70 percent of the population of many developing countries. With limited resources, they manage multiple activities in crop and animal production to spread risk and sustain their households. However, they are overlooked in public assistance programmes and services, which are often biased towards the bigger and better organized farmers. Numerous smallholders have less influence on policy, poor access to support services and technical inputs. They are generally seen to have a lower contribution to national food supply, and the cost of servicing them is considered as disproportionately high.

Small rural producers, particularly in remote areas, face difficulties in physical access to markets, placing constraints on the volumes and types of products they can produce, and restricting their potential to benefit from the growing demand in farm products. These problems are made worse by the lack of post-harvest facilities for highly perishable products, such as fish, and weak links to markets.

Farm ponds can provide for a number of benefits that can help smallholders in their quest for development. They provide for water storage, can be used to rear fish and other aquatic organisms like crustaceans, molluscs and plants, be integrated with other farm enterprises and improve and vary farm family diets. Further and importantly farm ponds can contribute to income and employment.

Aquaculture has been recognized as an important component of rural development strategies aimed at improving food supply and generating more income for poor farming households. To understand their circumstances better, the categorization 'commercial' and 'non-commercial' has been devised.

To clearly define the target group for support programmes, it is necessary to understand non-commercial and commercial fish farming. Experience over the past decades has shown that few farmers make the transformation from non-commercial to commercial farming.

Ponds add value to other farming activities: water from ponds can serve domestic and livestock water supplies as well as irrigation for high-value crops and vegetables. Farm ponds typically range from 30 m^2 to 500 m^2; size depends on the resources of the farmer and larger natural 'small water bodies', like small lakes, can also be considered as farm ponds.

Proximity of ponds to the farmer's house is vital for management and those which are fully integrated contribute most to the overall operation of the farm. Ponds offer the opportunity to rear fish and other aquatic organisms providing for diversification of farm enterprises and reduction of risk.

A well-managed integrated farm system comprising several enterprises including fish culture in ponds produces more products.

The pond serves as a water storage structure and, as such, becomes a key asset of smallholder farms, supplying water to crops and animals. Crops can be integrated with the ponds in a multiple use system. Wastes accumulated in pond muds can be spread over gardens. Fish ponds in irrigation schemes add value to the scheme and serve as a back-up source of water when water shortage occurs or supply is temporarily cut off. There are mutual benefits to integrated and associated farming systems, such as wastes from gardens being fed to fish and water from ponds for irrigating the crops.

Good fish production in farm ponds begins with the selection of type of pond system that is most compatible with prevailing conditions. It is necessary to be aware of the local traditional and agricultural calendars in order to organize farm pond labour as well as harvests. As populations grow, pressure on land and water resources increase and conflicts regarding access to and use of resources occur. These issues may also relate to the proximity of the farm pond to the farmer's core farm activities. Legal and social issues regarding small ponds include property rights, water rights and crop losses from poaching. Farm ponds isolated from the core farm often have unclear property rights and poaching can be a problem. Clearly it is easier to assist farmers having all activities in one location.

Types of Farm Pond Systems

Fish farming at the smallholder level could involve eight different types of farm pond systems:

1. Runoff ponds,
2. Ground-water ponds,
3. Borrow pit ponds,
4. Derivation ponds,
5. ponds using siphoned or pumped water,
6. Rice paddies,
7. Ponds integrated in irrigation schemes and
8. Small water bodies.

Ponds may vary in depth, but better production is obtained in ponds having a depth of at least 1.2 m and shallows at 0.3 m. This helps limit encroachment of aquatic vegetation as ponds with greater shallows may suffer from uncontrollable weed problems, which actually limit fish production.

Stream

A stream is a body of water with a current, confined within a bed and stream banks. Depending on its location or certain characteristics, a stream may be referred to as a *branch*, *brook*, *beck*, *burn*, *creek*, "crick", *gill* (occasionally *ghyll*), *kill*, *lick*, *mill race*, *rill*, *river*, *syke*, *bayou*, *rivulet*, *streamage*, *wash*, *run*, or *runnel*.

Streams are important as conduits in the water cycle, instruments in groundwater recharge, and corridors for fish and wildlife migration. The biological habitat in the immediate vicinity of a stream is called a riparian zone. Given the status of the ongoing Holocene extinction, streams play an important corridor role in connecting fragmented habitats and thus in conserving biodiversity. The study of streams and waterways in general is known as *surface hydrology* and is a core element of environmental geography.

Type

Brook: A stream smaller than a creek, especially one that is fed by a spring or seep. It is usually small and easily forded. A brook is characterised by its shallowness and its bed being composed primarily of rocks.

Creek:

- In North America, Australia and New Zealand, a small to medium sized natural stream. Sometimes navigable by motor craft and may be intermittent.
- In parts of New England, the UK and India, a tidal inlet, typically in a salt marsh or mangrove swamp, or between enclosed and drained former salt marshes or swamps (e.g. Port Creek separating Portsea Island from the mainland). In these cases, the stream is the tidal stream, the course of the seawater through the creek channel at low and high tide.

River: A large natural stream, which may be a waterway.

Runnel: The linear channel between the parallel ridges or bars on a shoreline beach or river floodplain, or between a bar and the shore. Also called a swale.

Tributary: A contributory stream, or a stream which does not reach the sea but joins another river (a parent river). Sometimes also called a branch or fork.

Coldwater Fish and Fisheries in the Indian Himalayas: Rivers and Streams

The Himalayas (Map) stretch for about 2500 km from west to east between Nanga Parbat (8126 m) in the west and Namcha Barwa (7756 m) in the east reaching their highest point in Mount Everest (8848 m). This mountain system is bordered in the west by the Hindu Kush and Karakoram mountains and in the north by the high Plateau of Tibet. The width from the south to the north varies between 200 and 400 km. Himalayas cover 594,400 km^2.

From south to north the Himalayas are divided into four parallel and longitudinal mountain belts of varying width, each having distinct physiographic features and its own geological history. These four divisions from south to north are known as the Siwaliks, the Lesser Himalaya, the Greater Himalaya and the Trans-Himalaya.

The Himalayas are drained by 19 major rivers, of which the Indus and the Brahmaputra are the longest, each having a mountain catchment of about 160,000 km^2.

Of the remaining 17 rivers five belong to the Indus system, of which the Beas and the Sutlej have a total catchment area of 80,000 km^2; nine (Ganga, Yamuna, Ram Ganga, Kali-Sharda, Karnali, Rapti, Gandak,

Bhagmati, Kosi) belong to the Ganga system, draining nearly 150,000 km^2; and three (Tista, Raidak, Manas) belong to the Brahmaputra system, draining another 110,000 km^2. Most of these rivers flow in deep valleys until they exit the mountains.

The deepening of the valleys and the elevation of the Himalayas took place simultaneously, resulting in the formation of mountain ranges with completely developed river systems. Different types of rocks and formations have resulted in rapids, cataracts, waterfalls, etc., which are common in the Himalayan chain of mountains.

In the physiography of the Himalayas the forest or vegetation cover forms an important element. Sub-tropical, temperate and alpine zones of vegetation are encountered as one moves from sub-Himalayan tracts to the snowy ranges.

The highest limit of forest growth in the more humid eastern Himalaya of Sikkim and Arunachal Pradesh is 4600-4900 m, and in drier Kumaon and Kashmir Himalaya the highest limit of tree growth is 4000-4200 m. Beyond this altitude trees and shrubs disappear and the mountains assume a rugged wind-swept and frost-bitten character.

The Himalayan rivers Indus, Ganga and Brahmaputra are among the largest rivers in the world. The mean annual discharge of 209,691.6 million m^{-3} of the Indus River is twice that of the Nile, and three times that of the Tigris and Euphrates combined. The principal rivers of the Indus system being snow-fed are all perennial but the flow in them varies enormously during the year. Floods occur in the rainy season (July-September), while the discharge is minimal during winter, although occasional freshets occur.

Unlike many other rivers of the world, the Indus and its tributaries receive all their water in the mountain region of their catchments. The flow characteristics vary from one river to another in the Indus system. In the main Indus, the discharges are lowest from mid-December to mid-February. The Jhelum and the Chenab have a winter minimum from mid-November to mid-January while in the Ravi, the Beas and the Sutlej the minimum is from mid-November to mid-February. The peaks occur from July to September. The Ganga has five source rivers, *viz.* the Bhagirathi, the Mandakini, the Alaknanda, the Dhauliganga and the Pindar.

The combined waters of the Alaknanda, the Dhauliganga and the Pindar join the Bhagirathi at Devaprayag to form the Ganga which emerges from the mountains at Rishikesh and flows down to Hardwar and to the northern plains of India.

The Brahmaputra, known as Yarlung Zangbo Jiang (Tsangpo) in Tibet, enters India under the name Dihang in Arunachal Pradesh. At its border with Assam it is joined by the Dibang and the Lohit, and from that point, turning west, the river is known as the Brahmaputra. The river carries clear water with little silt in Tibet, but on its entry into India the silt load increases. The river causes heavy floods during the south-west monsoon season.

Fish

The fish species distribution in the Himalayan streams depends on the flow rate, nature of substratum, water temperature and the availability of food. In torrential streams Sehgal identified several zones on the basis of dominant fish species and the hydrographical features:

(i) headwater zone inhabited by rheophilic species of loaches and catfishes (*Noemacheilus gracilis*, *N. stoliczkae* and *Glyptosternum reticulatum*);

(ii) large stream zone, formed by the joining of headwater streams, inhabited by *Diptychus maculatus* and

Noemacheilus spp. In the upper reaches or the most torrential reaches of this zone, rheophilic species of the snow trouts *Schizothoraichthys esocinus*, *S. progastus*, *Schizothorax richardsonii* and *Schizopygopsis stoliczkae* occur. The intermediate reaches of the large stream zones are frequented by *Schizothorax longipinnis*, *S. planifrons* and *S. micropogon*. The least rapid reaches of this zone are occupied by *Garra gotyla*, *Crossocheilus diplochilus*, *Labeo dero* and *L. dyocheilus*; (iii) slow moving meandering zone inhabited by a large number of cold- to eurythermal species such as *Barilius* spp., *Tor* spp. cat fishes, homalopterid fish (*Homaloptera* spp.) and snakeheads (*Channa* spp.).

The eastern Himalaya drained by the Brahmaputra has a greater diversity of coldwater fish than the western Himalayan drainage. For the whole Himalayas, 218 species are listed. The subsistence and commercial fisheries exploit carps (*Labeo* and *Tor* spp.), lesser barils (*Barilius* spp.), schizothoracines (*Schizothorax* and *Schizothoraichthys* spp.), garrids (*Garra* spp.) and sisorids (*Glyptothorax* and *Glyptosternum* spp.). The other genera are small-sized and of low economic value. The exotic brown trout (*Salmo trutta*) has established itself in some areas of the Himalayas.

The main factors which influence fish life in the Himalayan streams are:

(i) current velocity;
(ii) fluctuation in water discharge;
(iii) water temperature and dissolved oxygen level;
(iv) substratum;
(v) shelter from the current; and
(vi) food availability represented mostly by organisms clinging to and growing on rock and stone surfaces in fast current.

The Himalayan fish spend the major part of their life facing the current. This helps them in two ways: firstly, to maintain their upright position, and secondly, to make respiration easier. They have to open their mouths to take in water and boost the respiratory current. Shoals of lesser barils (*Barilius* spp.) are present in shallow pools (20-25 cm in depth). Loaches, which have colouration of their backs matching the background of sandy pools, swim rapidly in short bursts.

The need for shelter from the current has led to territoriality. Mahseers and schizothoracines chase intruders to defend the limited food resource and available shelter. Such a behaviour develops after the young fish emerges from the eggs laid in gravel. During winter months all size groups of mahseers and schizothoracines are present in pools when the water level is at its lowest and water is highly transparent. Such pools are present in the rivers Jhelum, Beas, Sutlej and Yamuna. Shoaling behaviour is seen in lesser barils and juvenile mahseers in shallow pools of the principal tributaries of the rivers Beas, Sutlej and Yamuna. This is one of the devices employed by these species to confuse predators. When a few fish are caught in a cast net, the rest disperse.

Water temperature is always an important limiting factor affecting geographical distribution and local occurrence within one water system. Cold stenothermic species such as the endemic schizothoracines (*Schizothoraichthys esocinus* and *Diptychus maculatus*) and exotic brown trout have an upper tolerance around 20°C. Carps, mahseers and lesser barils have a wider tolerance and even survive water temperatures over 25°C. Schizothoracines and brown trout remain active in the near-zero temperature which prevail in streams of the Lesser and Greater Himalaya during December and January.

Hailstorms and drought conditions in the Lesser Himalaya may cause adverse conditions. To cope with the steep fall in temperature in winter months schizothoracines migrate from headwaters to lower altitudes where they represent a sizeable part in fish catches in large rivers and their tributaries. The rise in temperature in Kashmir streams from near-freezing level to 10-17°C during May-June induces *S. richardsonii*, *S. longipinnis* and *S. curvifrons* to spawn.

In the Sutlej River, *S. richardsonii* starts upstream migration with the rise in water temperature during March. During the upstream migration the fish still finds itself in waters of low temperature of 8.0-9.5°C, owing to the steady influx of snow-melt water. This induces the species to migrate to and spawn in side streams, which receive warm ground water of 17.5-21.5°C. In the Ravi River system the fish spawn in May. In the upper Beas, however, the fish spawn only in July-August when the stream water temperature warms up to 16.5-18.5°C. In the same drainage *S. richardsonii* migrates downstream to the lowermost reaches where it spawns from October to December at 19.0 to 22.5°C. These observations indicate that in some schizothoracines multiple spawning is determined by temperatures and flow rates optimal for egg laying. The eggs are large-sized (3.0-4.0 mm diameter) and sticky in nature. They are laid in shallow pools (50-70 cm depth) and remain adhered to the substratum until the hatching of fry.

The fast-swimming species of mahseer, trout and schizothoracines expend much energy in maintaining an upright position in the turbulent and fast current. An attempt has been made to classify the coldwater fish on the basis of current velocity in the Jhelum basin of Kashmir. The frequent occurrence of spates has proved deleterious to breeding and propagation of coldwater fish. The scanty population as indicated by the low density of fish in the upper reaches of the Sutlej and Chenab rivers may result from the passage of these rivers through deep and narrow gorges, and the presence of cold glacier- and snow-melt water.

The fluctuating discharge of water and drying out of streams, leaving only isolated pools or no water at all, is another important matter. A study on seasonal fluctuation in river discharge in the Indus system indicates that the range of mean flow from October to March (winter months) represents only 8-10% of the total annual flow. There are also variations from year to year depending on the winter and monsoon precipitation. Reduction of torrential streams to stagnant pools exposes the fish to terrestrial predators and to depletion in dissolved oxygen concentrations, especially when autumn leaf fall takes place. However, due to low temperature, the level of dissolved oxygen may not fall below the optimum required by coldwater fish (7.0-8.0 mg l^{-1}). As soon as the flow is restored with spring rains and snow-melt water a rapid recolonisation of the stream takes place.

In the Himalayas, two zones can be distinguished. The rhithron zone is characterized by a monthly mean temperature of 17.3°C, high concentrations of dissolved oxygen (10.1 mg l^{-1}), fast current (0.9-1.8 m/sec), and turbulent waters. The substratum is rocks and boulders with sand and silt patches and with some pools. The fish of this zone are stenothermic, such as brown trout and snow trouts (*S. richardsonii*, *Schizothoraichthys esocinus* and *S. progastus*). This zone borders on the potamon zone which has a higher mean water temperature of 22.1°C, dissolved oxygen of 8.0 mg l^{-1}, and current velocity of 0.5-0.7 m/sec. The substratum consists of boulders, stones, gravel and patches of aquatic vegetation in the pools. The fish fauna is eurythermic or warm-stenothermic (*Labeo dero*, *Tor putitora* and *Garra gotyla*).

Chapter 4

Limnology and Habitat Features of the River Ganga

The Ganges, also Ganga, is a trans-boundary river of Asia which flows through India and Bangladesh. The 2,525 km (1,569 mi) river rises in the western Himalayas in the Indian state of Uttarakhand, and flows south and east through the Gangetic Plain of North India into Bangladesh, where it empties into the Bay of Bengal. It is the third largest river by discharge.

The Ganges is the most sacred river to Hindus. It is also a lifeline to millions of Indians who live along its course and depend on it for their daily needs. It is worshipped as the goddess *Ganga* in Hinduism. It has also been important historically, with many former provincial or imperial capitals (such as Pataliputra, Kannauj, Kara, Kashi, Allahabad, Murshidabad, Munger, Baharampur, Kampilya, and Kolkata) located on its banks.

The Ganges was ranked as the fifth most polluted river of the world in 2007. Pollution threatens not only humans, but also more than 140 fish species, 90 amphibian species and the endangered Ganges river dolphin. The Ganga Action Plan, an environmental initiative to clean up the river, has been a major failure thus far, due to corruption, lack of technical expertise, poor environmental planning, and lack of support from religious authorities.

Ecology and Environment

Human development, mostly agriculture, has replaced nearly all of the original natural vegetation of the Ganges basin. More than 95% of the upper Gangetic Plain has been degraded or converted to agriculture or urban areas. Only one large block of relatively intact habitat remains, running along the Himalayan foothills and including Rajaji National Park, Jim Corbett National Park, and Dudhwa National Park. As recently as the 16th and 17th centuries the upper Gangetic Plain harbored impressive populations of wild Asian elephants (*Elephas maximus*), tigers (*Panthera tigris*), Indian Rhinoceros (*Rhinoceros unicornis*), gaurs (*Bos gaurus*), barasinghas (*Rucervus duvaucelii*), sloth Bears (*Melursus ursinus*) and Indian lions.

In the 21st century there are few large wild animals, mostly deer, boars, wildcats, and small numbers of wolves, jackals, and foxes. Bengal tigers survive only in the Sundarbans area of the Ganges Delta. Crocodiles and barasingha are also found in the Sundarbans. The Sundarbands freshwater swamp ecoregion, however, is nearly extinct. Threatened mammals in the upper Gangetic Plain include the tiger, elephant, sloth bear, and chousingha (*Tetracerus quadricornis*).

Fish are found in all the major rivers of the Ganges basin, and are a vital food source for many people. In the Bengal area common fish include featherbacks (Notopteridae family), barbs (Cyprinidae), walking catfish (*Clarias batrachus*), gouramis (Anabantidae), and milkfish (*Chanos chanos*). The critically endangered Ganges shark (*Glyphis gangeticus*) is also found in the river and other places in south Asia.

Many types of birds are found throughout the basin, such as myna, parrots, crows, kites, partridges, and fowls. Ducks and snipes migrate across the Himalayas during the winter, attracted in large numbers to wetland areas. There are no endemic birds in the upper Gangetic Plain. The Great Indian Bustard (*Ardeotis nigriceps*) and Lesser Florican (*Sypheotides indicus*) are considered globally threatened.

The natural forest of the upper Gangetic Plain has been so thoroughly eliminated it is difficult to assign a natural vegetation type with certainty. There are a few small patches of forest left, and they suggest that much of the upper plains may have supported a tropical moist deciduous forest with sal (*Shorea robusta*) as a climax species.

The Ganges River itself supports the mugger crocodile (*Crocodylus palustris*) and the gharial (*Gavialis gangeticus*). The river's most famed fauna is the freshwater dolphin *Platanista gangetica gangetica*, the Ganges River dolphin, recently declared India's national aquatic animal.

A similar situation is found in the lower Gangetic Plain, which includes the lower Brahmaputra River. The lower plains contain more open forests, which tend to be dominated by *Bombax ceiba* in association with *Albizzia procera*, *Duabanga grandiflora*, and *Sterculia vilosa*.

There are early seral forest communities that would eventually become dominated by the climax species sal (*Shorea robusta*), if forest succession was allowed to proceed. In most places forests fail to reach climax conditions due to human causes. The forests of the lower Gangetic Plain, despite thousands of years of human settlement, remained largely intact until the early 20th century. Today only about 3% of the ecoregion is under natural forest and only one large block, south of Varanasi, remains. There are over forty protected areas in the ecoregion, but over half of these are less than 100 square kilometres (39 sq mi). The fauna of the lower Gangetic Plain is similar to the upper plains, with the addition of a number of other species such as the Smooth-coated Otter (*Lutrogale perspicillata*) and the Large Indian Civet (*Viverra zibetha*).

Ganges River Dolphin

The Ganges River Dolphin, which used to exist in large schools near to urban centres in both the Ganges and Brahmaputra rivers, is now seriously threatened by pollution and dam construction. Their numbers have now dwindled to a quarter of their numbers of fifteen years before, and they have become extinct in the Ganges's main tributaries. A recent survey by the World Wildlife Fund found only 3,000 left in the water catchment of both river systems.

The Ganges river dolphin is one of only four freshwater dolphins in the world. The other three are the baiji (*Lipotes vexillifer*) of the Yangtze River in China, now likely extinct, the bhulan of the Indus River in Pakistan, and the boto of the Amazon River in Brazil. There are several marine dolphins whose ranges include some freshwater habitats, but these four are the only dolphins who live only in freshwater rivers and lakes.

Physico-Chemical Characteristics:

i) Temperature range - 16.70C in January – 31.50C in June to sept.

ii) PH - 7.4 during June to August and Maximum 8.3 during January to May.

iii) Turbidity - 100 ppm in January; 1100-2170 ppm during July to September.

iv) Do2 - 5.0 to 10.5 ppm during January to February while in monsoon 2.00ppm (July-Sept.)

v) Co2 - 0.6 ppm -10.0ppm

vi) Chloride - 4.0 -35.4 ppm

vii) Phosphate - 0.05-021ppm

viii) Nitrates - 0.08-0.22ppm

ix) Silicates - 4.0-20.3ppm

x) Carbonates - 1.0 – 12.0 ppm

Common Phytoplanktons: Phytoplanktons are generally poor during the monsoon and autam months. Common phytoplanktons found in Ganga river system are –

(i) Members of Bacillariophyceae like Amphora, Asterionella, Cymbella, Navicula and Synedra etc.

(ii) members of Chlorophycace like, Chlorella, Closterium, Denticula, Pandorina and Spirogyea etc.

(ii) members of Myxophyceae like Anabaena, Nostoc, Oscillatoria etc.

Common zooplanktons: Rattulas, Rotaria, Keratella, Filuia, Notops, Monostyla etc.

Fisheries of Ganga river systems: The Ganga river system supports a large number of commercially important fish species including major carps (*Labeo. rohita: L.Calabasu, Catla catla and Cirrhinus mrigala*), minor carps (*Labeo fimbriatus; L.bata; Cirrhinus. reba*), catfishes (*Wallago. attu ; Mystus. aor; M.tengara, Clarias. batrachus; Heteropneustes fossilis*), cluipeiods, murrels (*Channa* species), feather backs (*Notopterus. notopterus; N.chitala*), mullets (*Mugil corsula*), fresh water eel (*Anguilla*) and prawns (*Macrobrachium malcolmsonii; Palaemon. Lamarii*). Apart from these fishes, the others like *Pangasius*; *silonia silondia*; *Gudusia chapra; Bagasius. bagasius; Eutropichthys. vacha* are also found in the river system.

The commercial fisheries in this zone are non-existing due to spares population, inaccessible terrain and poor communication between fishing grounds and landing centres.

The fish yield has been declined over the years due to 1) sandification of the river bed (upto Patna) which reduced the rivers productivity due to blanket effect, (2) marked reduction in the water volume on account of increase sedimentation, (3) increased water abstraction and (4) irrational fishing.

In spite of this, the Ganga river system is contributing nearly about 89.5% of the total fish seed correlation of India.

The effects of climate change: The Tibetan Plateau contains the world's third-largest store of ice. Qin Dahe, the former head of the

China Meteorological Administration, said that the recent fast pace of melting and warmer temperatures will be good for agriculture and tourism in the short term; but issued a strong warning:

> *"Temperatures are rising four times faster than elsewhere in China, and the Tibetan glaciers are retreating at a higher speed than in any other part of the world.... In the short term, this will cause lakes to expand and bring floods and mudflows... In the long run, the glaciers are vital lifelines for Asian rivers, including the Indus and the Ganges. Once they vanish, water supplies in those regions will be in peril."*

In 2007, the Intergovernmental Panel on Climate Change (IPCC), in its Fourth Report, stated that the Himalayan glaciers which feed the river, were at risk of melting by 2035. The IPCC has now withdrawn that prediction, as the original source admitted that it was speculative and the cited source was not a peer reviewed finding. In its statement, the IPCC stands by its general findings relating to the Himalayan glaciers being at risk from global warming (with consequent risks to water flow into the Gangetic basin).

Study of Habitat and Fish Biodiversity of River Kopili in Assam

Kopili River is an interstate river in Northeast India that flows through the states of Meghalaya and Assam and is the largest south bank tributary of the Brahmaputra in Assam.The Kopili originates in the Meghalaya plateau and flows through southern Assam before its confluence with the Brahmaputra. In Assam it drains the districts of Karbi Anglong, Dima Hasao, Kamrup and Nagaon. The river flows for a total length of 290 kilometres (180 mi) and has a catchment area of 16,420 square kilometres (6,340 sq mi). It is noted for several spectacular waterfalls along its course which has several deep gorges and rapids in the 120 kilometres (75 mi) of its flow before debouching into the plains at Nagaon district.

Waterworks

Completed in 1975, the Kopili Flow Irrigation Scheme in Kamrup district irrigates 1,300 hectares (3,200 acres) of land across 14 revenue villages and facilitates paddy cultivation. The Kopili Hydro Electric Project, located across the districts of North Cachar Hills in Assam and Jaintia Hills in Meghalaya and run by the North Eastern Electric Power Corporation, consists of the Khandong and Umrongso dams and their reservoirs and three power houses that have a total installed capacity of 275 MW.

Environmental Issues

The Kopili has as many as 54 species of fish in it. Unscientific opencast coal mining in the Kopili's upper reaches in Meghalaya has led to acidification of the river which has in turn left part of the river's course biologically dead, making the water unfit for human consumption and has led to frequent outages at the Kopili Hydro Electric Project's dams.

Physico-chemical Characteristics of Water

Wetlands are currently degraded by both natural and anthropogenic activities, which deteriorate their quality, and push them to the brink of extinction in the process of unplanned development, giving rise to the need for suitable conservation strategies. Unfortunately, over the years, less attention has been given to wetland losses world over, including Bangalore. The degradation of the wetlands has altered their functions, affecting the ecological balance. The objectives of carrying out the physico-chemical and biological analyses of water bodies are as follows:

1. Generally, lake functions directly relate to their physical, chemical and biological integrity. Water quality evaluation for wetlands leads to information about their misuse by indicating the pollution status. Gale et al. (1993) define water quality objectives as the overall direction and purpose of the project, and furthermore define goals as milestones to be met during the course of a project. Since the quality of aquatic life depends on the water quality, a thorough assessment of the water quality is an integral part of wetland evaluation. The assessment of the chemical criteria of the waterbody helps in:
    - Evaluating the chemicals that cause toxicity to aquatic life.
    - Studying the long-term effects on the ecosystem.
    - Conducting the status and monitoring of wetland resources by studying their physico-chemical and biological parameters.
2. Designate uses that protect the structure and function of wetlands for protection of fish, birds, wildlife, and recreation. The baseline values attached to wetlands in terms of designating the viable usage of these water bodies based on established standards protecting their functions is also attempted.
3. Analyse the qualitative and quantitative aspects of plankton population of the water bodies. The biological integrity of the

wetlands is the driving force for their sound ecological functioning. Wetlands that support a vast diversity of fish, birds, mammals, etc., depend directly on it by supporting vast and diverse forms of plankton providing a nutrient base and a complex food-web. In many cases, planktons also act as the biological indicators of pollution being very sensitive to changes in water quality.

Methodology

Water is a dynamic medium and its quality varies spatially and temporally. In order to characterise any water body, studies on the major components, hydrology, physico-chemical and biological characteristics, should be carried out.

Hydrological Features

A thorough knowledge of the hydrological properties of the water body must be acquired before an effective water quality monitoring system is established. Each of the inland waterbody is characterised by unique hydrological features such as:

- *Rivers:* characterised by uni-directional current with relatively high average velocity (0.1 – 1.0 m/S). In general, thorough and continuous vertical mixing is achieved in rivers due to the prevailing currents and turbulence.
- *Lakes:* characterised by low, average current (0.001 to 0.01 m/S) giving higher residential time for water. Currents within a lake are multi-directional with mixing regulated by the climatic conditions and lake depth.
- *Ground water*: characterised by a steady flow pattern both in direction and speed that is largely governed by the porosity of the geological material as a result of which the mixing is poor.
- *Reservoirs*: intermediate between rivers and lakes.

Large variations in the water residence time occur in different types of inland water bodies. The hydrodynamic characteristics of each type of water body are highly dependent on the size of the waterbody, climatic conditions and the drainage pattern associated with it. Ground water greatly depends on the recharge regime i.e. infiltration through unsaturated aquifer zone, that allows renewal of the groundwater.

Physical, Chemical and Biological Properties

The physical and chemical properties of a freshwater body are characteristic of the climatic, geochemical, geomorphological and pollution conditions (largely) prevailing in the drainage basin and the underlying aquifer.

EXPERIMENT	PRESERVATIVE	Max. holding time
BOD	Cool, 4° C	4 hours
Calcium	Cool, 4° C	7 days
Chloride	Cool, 4° C	7 days
COD	Cool, 4° C	24 hours
Dissolved Oxygen*	Fix on site	6 hours
Fluoride	Cool, 4° C	7 days
Magnesium	Cool, 4° C	7 days
Nitrate + Nitrite	Cool, 4° C	24 hours
PH	None	6 hours
Phosphorus*		
Dissolved	Filter on site using 0.45μm filter	24 hours
Inorganic	Cool, 4° C	24 hours
Ortho	Cool, 4° C	24 hours
Total	Cool, 4° C	1 month
Potassium	Cool, 4° C	7 days
Specific conductance	Cool, 4° C	24 hours
Sodium	Cool, 4° C	7 days
Heavy metals		
Cadmium	2 ml conc. Nitric acid/L sample	6 months
Chromium	2 ml conc. Nitric acid/L sample	6 months
Copper	2 ml conc. Nitric acid/L sample	6 months
Iron	2 ml conc. Nitric acid/L sample	6 months
Lead	2 ml conc. Nitric acid/L sample	6 months
Nickel	2 ml conc. Nitric acid/L sample	6 months
Zinc	2 ml conc. Nitric acid/L sample	6 months

(Source: Analytical Methods Manual, Water quality branch, Environment Canada, 1981)

The biota in the surface water is governed entirely by various environmental conditions that determine the selection of species as the physiological performance of the individual organisms. The primary production of organic matter, in the form of phytoplankton and macrophytes is more intense in lakes and reservoirs than in rivers. In contrast to the chemical quality of water bodies, which can be measured by suitable analytical methods, biological quality is a combination of both qualitative and quantitative characterisation. This can be carried out in two levels:

- Response of individual species to changes in its environment.
- Response of biological communities to changes in their environment.

Analyses of Physical, Chemical and Biological Parameters

The parameters analysed to assess the water quality are broadly divided into:

- *Physical parameters:* Colour, Temperature, Transparency, Turbidity and Odour.
- *Chemical parameters:* pH, Electrical Conductivity (E.C), Total Solids (TS), Total Dissolved Solids (TDS), Total Suspended Solids (TSS), Total Hardness, Calcium Hardness, Magnesium Hardness, Nitrates, Phosphates, Sulphates, Chlorides, Dissolved Oxygen (D.O), Biological Oxygen Demand (BOD), Chemical Oxygen Demand (COD), Fluorides, Free Carbon-di-oxide, Potassium and Sodium.
- *Heavy metals:* Lead, Copper, Nickel, Iron, Chromium, Cadmium and Zinc.
- *Biological parameters:* The biological parameters involved the qualitative analyses of planktons (zooplankton and phytoplankton).

Field measurement: The field parameters measured include pH, conductivity, dissolved oxygen, temperature and transparency.

Physical Parameters

Colour

In natural water, colour is due to the presence of humic acids, fulvic acids, metallic ions, suspended matter, plankton, weeds and industrial effluents. Colour is removed to make water suitable for general and industrial applications and is determined by visual comparison of the sample with distilled water.

Visual comparison: About 20ml of the sample and 20ml of distilled water were taken in two separate wide mouthed test tubes. The results were tabulated (as clear, greenish, greyish, brownish, blackish, etc) by comparing the colour of the sample with distilled water.

Temperature

Impinging solar radiation and atmospheric temperature brings about spatial and temporal changes in temperature, setting up convection currents and thermal stratification. Temperature plays a very important role in wetland dynamism affecting the various parameters such as alkalinity, salinity, dissolved oxygen, electrical conductivity etc. In an aquatic system, these parameters affect the chemical and biological reactions such as solubility of oxygen, carbon-di-oxide-carbonate-bicarbonate equilibrium, increase in metabolic rate and physiological reactions of organisms, etc. Water temperature is important in relation to fish life. The temperature of drinking water has an influence on its taste.

Apparatus required: Thermometer- 0.1° C division.

Procedure: Temperature measurement is made by taking a portion of the water sample (about 1litre) and immersing the thermometer into it for a sufficient period of time (till the reading stabilizes) and the reading is taken, expressed as °C.

Transparency (Light Penetration)

Solar radiation is the major source of light energy in an aquatic system, governing the primary productivity. Transparency is a characteristic of water that varies with the combined effect of colour and turbidity. It measures the light penetrating through the water body and is determined using Secchi disc.

Apparatus required: Secchi disc, a metallic disc of 20cm diameter with four quadrats of alternate black and white on the upper surface. The disc with centrally placed weight at the lower surface, is suspended with a graduated cord at the centre.

Procedure: Transparency is measured by gradually lowering the Secchi disc at respective sampling points. The depth at which it disappears in the water (X_1) and reappears (X_2) is noted. The transparency of the water body is computed as follows:

Transparency (Secchi Disc Transparency) = $(X_1 + X_2)/2$

Where, X_1 = Depth at which Secchi disc disappears

X_2 = Depth at which Secchi disc reappears

Turbidity

Turbidity is an expression of optical property; wherein light is scattered by suspended particles present in water (Tyndall effect) and is measured using a nephelometer. Suspended and colloidal matter such as clay, silt, finely divided organic and inorganic matter; plankton and other microscopic organisms cause turbidity in water. Turbidity affects light scattering, absorption properties and aesthetic appearance in a water body. Increase in the intensity of scattered light results in higher values of turbidity.

Apparatus required: Nephelometer (It detects scattered light at 90° to the incident beam of light. It consists of a light source for illuminating the sample. One or more photoelectric detectors with a display unit indicate the intensity of light scattered at 90° to the path of incident light.), sample cells, lab-glass wares and Monopan balance.

Principle: Nephelometric measurement is based on comparison of the intensity of scattered light of the sample with the intensity of light scattered by a standard reference suspension (Formazin polymer) under similar conditions.

Reagents:

- Distilled water and Stock primary Formazin suspension:
  - *Solution* 1: 1.0 g Hydrazine sulphate is dissolved in 100ml of distilled water.
  - *Solution* 2: 10.0g of Hexamethylenetetramine is dissolved in distilled water and made up to 100ml in a volumetric flask.
- *Stock Turbidity Suspension*: 5ml of solutions 1 and 2 are mixed in a volumetric flask and allowed to stand for 24 hrs at about 25° C (±3° C) and diluted to 1000ml with distilled water to give a 400 NTU suspension.
- Standard Turbidity Suspension: 10ml of the stock solution is diluted to 100ml with distilled water to give a standard solution of 40NTU.

Procedure: The nephelometer is calibrated using distilled water (Zero NTU) and a standard turbidity suspension of 40NTU. The thoroughly shaken sample is taken in the nephelometric tube and the value is recorded.

Turbidity (NTU) = (Nephelometer readings) (Dilution factor*)

* If the turbidity of the sample is more than 40 NTU, then the sample is diluted and the dilution factor is accounted in final calculations.

Chemical Parameters

pH:

The effect of pH on the chemical and biological properties of liquids makes its determination very important. It is one of the most important parameter in water chemistry and is defined as -log [H^+], and measured as intensity of acidity or alkalinity on a scale ranging from 0-14. If free H^+ are more it is expressed acidic (i.e. pH<7), while more OH^- ions is expressed as alkaline (i.e. pH> 7).

In natural waters pH is governed by the equilibrium between carbon dioxide/bicarbonate/carbonate ions and ranges between 4.5 and 8.5 although mostly basic. It tends to increase during day largely due to the photosynthetic activity (consumption of carbon-di-oxide) and decreases during night due to respiratory activity. Waste water and polluted natural waters have pH values lower or higher than 7 based on the nature of the pollutant.

* The colourimetric indicator method can be used only for approximate pH values.

Apparatus required:

- pH indicator (BDH) method: BDH Indicator (Universal Indicator) and test tubes.
- Electrometric method: Glass electrode, reference electrode (mercury/calomel or silver/silver chloride) and pH meter.

Procedure:

- *Colourimetric method:* About 10ml of the sample is taken in a wide mouth test tube, 0.2ml of BDH indicator is added, and shaken gently. The colour developed is matched with the chart and the pH noted.
- *Electrometric method*: The pH is determined by measuring the Electro Motive Force (E.M.F) of a cell comprising an indicator electrode (an electrode responsive to hydrogen ions such as a glass electrode) immersed in the test solution and the reference electrode (usually a mercury/calomel electrode).

  Contact between the test solution and the reference electrode is usually got by means of a liquid junction, which forms a part of reference electrode. E.M.F of this cell is measured with pH meter, that is a high impedance voltmeter calibrated in terms of pH. The electrode is allowed to stand for 2 minutes to stabilize before taking reading for reproducible results (at least ±0.1 pH units).

Electrical Conductivity

Conductivity (specific conductance) is the numerical expression of the water's ability to conduct an electric current. It is measured in micro Siemens per cm and depends on the total concentration, mobility, valence and the temperature of the solution of ions. Electrolytes in a solution disassociate into positive (cations) and negative (anions) ions and impart conductivity.

Most dissolved inorganic substances are in the ionised form in water and contribute to conductance.

The conductance of the samples gives rapid and practical estimate of the variation in dissolved mineral content of the water supply.

Conductance is defined as the reciprocal of the resistance involved and expressed as mho or Siemen (s).

$$G = \frac{1}{R}$$

G – Conductance (mho or Siemens) and R - Resistance

Apparatus required: Conductivity meter

Procedure: The electrode of the conductivity meter is dipped into the sample, and the readings are noted for stable value shown as mS/cm.

TOTAL SOLIDS:

Total solids is the term applied to the material residue left in the vessel after evaporation of the sample and its subsequent drying in an oven at a temperature of 103-105°C. Total solids include Total Suspended Solids (TSS) and Total Dissolved Solids (TDS).

Principle: A known volume (50 ml) of well-mixed sample is evaporated in a pre-weighed dish and dried to constant weight in an oven at 103-105° C. The increase in weight over that of the empty dish gives the total solids.

Apparatus: Evaporating dishes-100ml porcelain dish, steam bath, drying oven, desiccator, Monopan balance and measuring jars.

Procedure: A known volume of the well-mixed sample (50ml) is measured into a pre-weighed dish and evaporated to dryness at 103°C on a steam bath.

The evaporated sample is dried in an oven for about an hour at 103-105° C and cooled in a desiccator and recorded for constant weight.

Calculation:

$$\text{Total solids (mg/L)} = \frac{(W1\text{-}W2)\ (1000)}{\text{Sample volume (ml)}}$$

W_1 = Weight of dried residue + dish

W_2 = Weight of empty dish

Total Dissolved Solids

Dissolved solids are solids that are in dissolved state in solution. Waters with high dissolved solids generally are of inferior palatability and may induce an unfavourable physiological reaction in the transient consumer.

Principle: The difference in the weight of total solids and the total suspended solids expressed in the same units gives the total dissolved solids.

Apparatus: Glass-fibre filter disks, membrane filter funnel, filtration apparatus, suction flask and pump, drying oven and Grooch crucible.

Procedure: The difference in the weights of Total Solids (W_1) and Total Suspended Solids (W_2) expressed in the same units gives Total Dissolved Solids (TDS).

Calculation:

$$\text{Total Dissolved Solids (mg/L)} = \frac{(W_1\text{-}W_2) \times 1000}{\text{Sample volume (ml)}}$$

W_1 = Weight of total solids + dish

W_2 = Weight of total suspended solids

Total Suspended Solids

Suspended solids are the portions of solids that are retained on a filter of standard specified size (generally 2.0 μ) under specific conditions. Water with high-suspended solids is unsatisfactory for bathing, industrial and other purposes.

Principle: A well – mixed sample is filtered through a weighed standard glass fibre filter and the residue that is retained on the filter is dried to a constant weight at 103-105 °C. The increase in the weight of the filter determines the total suspended solids.

Apparatus: Porcelain dish (100ml capacity), glass fibre filter disk, suction pump and flask, measuring jar, membrane filter funnel, oven and filtration apparatus.

Procedure: The known volume of vigorously shaken sample (50ml) is filtered into a pre-weighed glass fibre filter disk fitted to suction pump, and washed successively with distilled water. The filter is carefully removed from the filtration apparatus and dried for an hour at 103-105 °C in an oven, cooled in dessicator and weighed for constant weight.

Calculation:

$$\text{Total Suspended Solids (mg/L)} = \frac{(W_1\text{-}W_2)\ (1000)}{\text{Sample volume (ml)}}$$

W_1 = Weight of dried glass fibre filter + residue

W_2 = Weight of glass fibre filter disk before filtering

Total Hardness

Hardness is predominantly caused by divalent cations such as calcium, magnesium, alkaline earth metal such as iron, manganese, strontium, etc. The total hardness is defined as the sum of calcium and magnesium concentrations, both expressed as $CaCO_3$ in mg/L. Carbonates and bicarbonates of calcium and magnesium cause temporary hardness. Sulphates and chlorides cause permanent hardness.

Hardness Chart (for drinking water):

Soft	0 – 60 mg/L
Medium	60 –120 mg/L
Hard	120 - 180 mg/L
Very Hard	> 180 mg/L

Principle: In alkaline conditions EDTA (Ethylene-diamine tetra acetic acid) and its sodium salts react with cations forming a soluble chelated complex when added to a solution. If a small amount of dye such as Eriochrome black-T is added to an aqueous solution containing calcium and magnesium ions at alkaline pH of 10.0 ± 0.1, it forms wine red colour. When EDTA is added as a titrant, all the calcium and magnesium ions in the solution get complexed resulting in a sharp colour change from wine red to blue, marking the end point of the titration. Hardness of water prevents lather formation with soap rendering the water unsuitable for bathing and washing. It forms scales in boilers, making it unsuitable for industrial usage. At higher pH>12.0, Mg^{++} ion precipitates with only Ca^{++} in solution. At this pH, murexide indicator forms a pink colour with Ca^{++} ion. When EDTA is added Ca^{++} gets complexed resulting in a change from pink to purple indicating end point of the reaction.

Apparatus required: Lab glassware-burette, pipette, conical flask, beakers etc.

Reagents:

- *Buffer solution:* 16.9 g of ammonium chloride and 1.25g of magnesium salt of EDTA is dissolved in 143ml of concentrated ammonium hydroxide and diluted to 250ml with distilled water.
- *Eriochrome black-T indicator:* 0.5 g of Eriochrome black-T indicator is dissolved in 100g of triethanolamine.
- *Standard EDTA titrant:* 0.01M or Ng AR grade EDTA is dissolved in distilled water and diluted to 1000ml and is standardised against standard calcium solution, 1ml = 1mg $CaCO_3$.
- *Standard Calcium Solution:* 1.0g of AR grade $CaCO_3$ is weighed into a 250ml conical flask, to which 1+1 HCl is added till all $CaCO_3$ is dissolved completely. 200ml of distilled water is added and boiled to expel carbon-di-oxide, and diluted to 1000ml. 1ml = 1mg $CaCO_3$.

Procedure: Exactly 50ml of the well-mixed sample is pipetted into a conical flask, to which 1ml of ammonium buffer and 2-3 drops of Eriochrome black -T indicator is added. The mixture is titrated against standard 0.01M EDTA until the wine red colour of the solution turns pale blue at the end point.

Calculation:

$$\text{Total hardness (mg/L)} = \frac{(T)\,(1000)}{V}$$

Where, T = Volume of titrant

V = Volume of sample

Calcium Hardness

The presence of calcium (fifth most abundant) in water results from passage through or over deposits of limestone, dolomite, gypsum and such other calcium bearing rocks. Calcium contributes to the total hardness of water and is an important micro-nutrient in aquatic environment and is especially needed in large quantities by molluscs and vertebrates. It is measured by EDTA titrimetric method. Small concentration of calcium carbonate prevents corrosion of metal pipes by laying down a protective coating. But increased amount of calcium precipitates on heating to form harmful scales in boilers, pipes and utensils.

Principle: When EDTA (Ethylene-diamine tetra acetic acid) is added to the water containing calcium and magnesium, it combines first with calcium. Calcium can be determined directly with EDTA when pH is made sufficiently high such that the magnesium is largely precipitated as hydroxyl compound (by adding NaOH and iso-propyl alcohol). When murexide indicator is added to the solution containing calcium, all the calcium gets complexed by the EDTA at pH 12-13. The end point is indicated from a colour change from pink to purple.

Apparatus required: Burettes, pipette, conical flask, beakers and droppers.

Reagents:

- *Sodium hydroxide (8%):* 8g of sodium hydroxide is dissolved in 100ml of distilled water.
- *Murexide indicator (ammonium purpurate):* 0.2 g of murexide is ground well with 100g of sodium chloride thoroughly.
- *Standard EDTA titrant, 0.01M:* 3.723 g of EDTA (disodium salt) is dissolved in distilled water and made up to 100ml with the same.

Procedure: A known volume (50ml) of the sample is pipetted into a clean conical flask, to which 1ml of sodium hydroxide and 1ml of iso-propyl alcohol is added. A pinch of murexide indicator is added to this mixture and titrated against EDTA until the pink colour turns purple.

Calculation:

$$\text{Calcium as Ca (mg/L)} = \frac{T \times 400.5 \times 1.05}{\text{Sample taken, ml}}$$

Where, T= volume of titrant, ml

$$\text{Calcium hardness (mg/L as } CaCO_3\text{)} = \frac{T \times 1000 \times 1.05}{\text{Sample taken, ml}}$$

Magnesium Hardness

Magnesium is a relatively abundant element in the earth's crust, ranking eighth in abundance among the elements. It is found in all natural waters and its source lies in rocks, generally present in lower concentration than calcium. It is also an important element contributing to hardness and a necessary constituent of chlorophyll. Its concentration greater than 125 mg/L can influence cathartic and diuretic actions.

Principle: Magnesium hardness can be calculated from the determined total hardness and calcium hardness.

Calculation:

Magnesium = (T - C) x 0.243
(as mg/L)

where, T = Total hardness mg\L (as $CaCO_3$)

C = Calcium hardness mg\L (as $CaCO_3$)

High concentration of magnesium proves to be diuretic and laxative, and reduces the utility of water for domestic use while a concentration above 500 mg/L imparts an unpleasant taste to water and renders it unfit for drinking. Chemical softening, reverse osmosis and electro dialysis or ion exchange reduces the magnesium hardness to acceptable levels.

Nitrates

Nitrates are the most oxidized forms of nitrogen and the end product of the aerobic decomposition of organic nitrogenous matter. The significant sources of nitrates are chemical fertilizers from cultivated lands, drainage from livestock feeds, as well as domestic and industrial sources. Natural waters in their unpolluted state contain only minute quantities of nitrates. The stimulation of plant growth by nitrates may result in eutrophication, especially due to algae. The subsequent death and decay of plants produces secondary pollution. Nitrates are most important for biological oxidation of nitrogenous organic matter. Certain nitrogen fixing bacteria and algae have the capacity to fix molecular nitrogen in nitrates. The main source of polluting nitrates is domestic sewage. Nitrates may find their way into ground water through leaching from soil and at times by contamination. They can be measured by the phenoldisulphonic method.

Principle: Nitrates react with phenoldisulphonic acid and produce a nitrate derivative, which in alkaline solution develops yellow colour due to rearrangement of its structure. The colour produced is directly proportional to the concentration of nitrates present in the sample.

Apparatus required: Nessler's tube, pipettes, beakers, spectrophotometer, cuvettes, measuring jar and hot water bath.

Reagents:

- Phenol disulphonic acid: 25 g of phenol is dissolved in 150 ml of concentrated sulphuric acid, to which 85ml of sulphuric acid is further added and heated for about 90 min on a water bath and stored in dark bottles upon cooling.
- Sodium hydroxide: About 50g of sodium hydroxide is dissolved in 150-200 ml of water and cooled.

- Conc. Ammonium hydroxide
- Nitrate solution:
- Stock nitrate solution: 721.8 mg (0.722g) of AR potassium nitrate is dissolved in distilled water and made up to 100ml for stock solution.
- Standard nitrate solution: Standard nitrate solution is prepared by evaporating 50ml of the stock solution to dryness in the water bath. The obtained residue is dissolved in 2ml of phenol disulfonic acid and diluted to 500ml, to give 1ml = 10 mg. The solution of various strengths ranging from 0.0 (blank) to 1.0 mg/L at the intervals of 0.2 mg/L is prepared by diluting stock solution with distilled water.

Procedure: A known volume (50ml) of the sample is pipetted into a porcelain dish and evaporated to dryness on a hot water bath. 2ml of phenol disulphonic acid is added to dissolve the residue by constant stirring with a glass rod. Concentrated solution of sodium hydroxide or conc. ammonium hydroxide and distilled water is added with stirring to make it alkaline. This is filtered into a Nessler's tube and made up to 50ml with distilled water. The absorbance is read at 410nm using a spectrophotometer after the development of colour. The standard graph is plotted by taking concentration along X-axis and the spectrophotometric readings (absorbance) along Y-axis. The value of nitrate is found by comparing absorbance of sample with the standard curve and expressed in mg/L.

Calculation:

$$\text{Nitrates (as mg/L)} = \frac{\text{Absorbance of sample X Conc. of Std X 1000}}{\text{Absorbance of Std. X Sample taken}}$$

The high concentration of nitrate in water is indicative of pollution.

Phosphates

Phosphates occur in natural or wastewaters as orthophosphates, condensed phosphates and naturally found phosphates. Their presence in water is due to detergents, used boiler waters, fertilizers and biological processes. They occur in solution in particles or as detritus. They are essential for the growth of organisms and a nutrient that limits the primary productivity of the water body. Inorganic phosphorus plays a dynamic role in aquatic ecosystems; when present in low concentration is one of the most important nutrients, but in excess along with nitrates and potassium, causes algal blooms. It is calculated by the stannous chloride method.

Principle: In acidic conditions orthophosphate reacts with ammonium molybdate forming Molybdophosphoric acid, reduced further to molybdenum blue by stannous chloride.

The intensity of the blue colour is directly proportional to the concentration of phosphate. The absorbance is noted at 690nm using spectrophotometer.

Apparatus required: Spectrophotometer, lab glassware, hot plate and Nessler's tube.

Reagents:

- Ammonium molybdate reagent: 25g ammonium molybdate is dissolved in 175ml distilled water. 280ml concentrated sulphuric acid is added to 400ml distilled water and cooled. Molybdate solution is added and the mixture diluted to 1000ml.
- Stannous chloride reagent: 2.5g fresh stannous chloride is dissolved in 100ml glycerol, heated on water bath and stirred with the glass rod to hasten dissolution.
- Standard phosphate solution: 219.5 mg of dried AR potassium hydrogen phosphate is dissolved in distilled water and made up to 1000ml, where 1ml = 50.0 mg. of phosphate. 10ml of the stock solution is made up to 1000ml to give 1ml = 0.05 mg. Standards of strength ranging from 0 (blank) to 0.05mg/L at intervals of 0.01mg is prepared by diluting the stock with distilled water.

Procedure: To 50ml of the filtered sample, 4ml of ammonium molybdate reagent and about 4-5 drops of stannous chloride reagent is added. After about 10 min but before 12 min, the colour developed is measured photometrically at 690nm and calibration curve is prepared. A reagent blank is always run with same treatment with distilled water as sample. The value of phosphate is obtained by comparing absorbance of sample with the standard curve and expressed as mg/L.

Calculation:

$$\text{Phosphates (as mg/L)} = \frac{\text{Absorbance of sample X Conc. of Std X 1000}}{\text{Absorbance of Std. X Sample taken}}$$

High phosphorus content causes increased algal growth till nitrogen becomes limiting, although blue green algae continues to dominate because of its ability to utilize molecular nitrogen. Besides sedimentation, high uptake by phytoplankton is one of the reasons for fast depletion of phosphorus in water.

Sulphates

Sulphates are found appreciably in all natural waters, particularly those with high salt content. Besides industrial pollution and domestic sewage, biological oxidation of reduced sulphur species also add to sulphate content. Soluble in water, it imparts hardness with other cations. Sulphate causes scaling in industrial water supplies, and odour and corrosion problems due to its reduction to hydrogen sulphide. It can be calculated by turbidometric method.

Principle: Sulphate ions are precipitated in acetic acid medium with barium chloride to form barium sulphate crystals of uniform size. The scattering of light by the precipitated suspension (barium sulphate) is measured by a Nephelometer and the concentration is recorded.

Apparatus required: Nephelometer, magnetic stirrer, Nessler's tubes and lab glassware.

Reagents:

- Conditioning reagent: 50 ml of glycerol was mixed in a solution containing 30 ml of conc. hydro chloric acid, 300ml distilled water (10% HCl), 100 ml of 95% ethyl alcohol or isopropyl alcohol and 75g NaCl.
- Barium Chloride
- Standard sulphate solution: 147.9mg of AR grade sodium sulphate was dissolved in distilled water and made up to 1000ml, to give 1ml = 100mg sulphate.

Procedure: 100ml of the sample is filtered into a Nessler's tube containing 5ml of conditioning reagent. About 0.2g of barium chloride crystals is added with continued stirring. A working standard is prepared by taking 1ml of the standard, 5ml of conditioning reagent and made up to 100ml, to give 100 NTU. The turbidity developed by the sample and the standards are measured using a Nephelometer and the results are tabulated.

Calculation:

Sulphate = (Nephlometric reading) (0.4) (Dilution Factor)
(as mg/L)

Chlorides

The presence of chlorides in natural waters can mainly be attributed to dissolution of salt deposits in the form of ions (Cl^-). Otherwise, high concentrations may indicate pollution by sewage, industrial wastes, intrusion of seawater or other saline water. It is the major form of inorganic anions in water for aquatic life. High chloride content has a deleterious effect on metallic pipes and structures, as well as agricultural plants. They are calculated by Argentometric method.

Principle: In alkaline or neutral solution, potassium chromate indicates the endpoint of the silver nitrate titration of chlorides. Silver chloride is quantitatively precipitated before the red silver chromate is formed.

Apparatus required: Lab glassware.

Reagents:

- Potassium chromate indicator solution: 50g of potassium chromate is dissolved in minimum amount of distilled water and silver nitrate is added drop wise till a red precipitate is formed. The mixture is allowed to stand for about 12 hours and diluted to 1000ml with distilled water.
- Silver nitrate solution (0.014N): 2.395g of silver nitrate is dissolved in distilled water and made up to 1000ml.

Procedure: A known volume of filtered sample (50ml) is taken in a conical flask, to which about 0.5ml of potassium chromate indicator is added and titrated against standard silver nitrate till silver dichromate ($AgCrO_4$) starts precipitating.

Calculation:

$$\text{Chlorides } (Cl^-) = \frac{(A\text{-}B)\ (N)\ (35.45)}{\text{Sample taken in ml}}$$

Where,

A - Volume of silver nitrate consumed by the sample

B - Volume of silver nitrate consumed by the blank

N - Normality of silver nitrate (Standard methods, APHA, 16th edn, pp 286-88)

Dissolved Oxygen

Oxygen dissolved in water is a very important parameter in water analysis as it serves as an indicator of the physical, chemical and biological activities of the water body. The two main sources of dissolved oxygen are diffusion of oxygen from the air and photosynthetic activity. Diffusion of oxygen from the air into water depends on the solubility of oxygen, and is influenced by many other factors like water movement, temperature, salinity, etc. Photosynthesis, a biological phenomenon carried out by the autotrophs, depends on the plankton population, light condition, gases, etc. Oxygen is considered to be the major limiting factor in water bodies with organic materials. Dissolved oxygen is calculated by many methods.

Method 1: Membrane electrode method

Principle: The membrane electrode has a sensing element protected by an oxygen-permeable plastic membrane that serves as a diffusion barrier against impurities.

Under steady conditions the electric current read is directly proportional to the D.O concentrations (electric current is directly proportional to the activity of molecular oxygen).

Apparatus required: Oxygen-sensitive membrane electrode and lab glassware.

Procedure: The calibrations are carried out following the manufacturer's calibration procedure. The electrode is dipped into the sample, and the reading noted.

Method 2: Winkler's method

Principle: Oxygen present in the sample oxidizes the dispersed divalent manganous hydroxide to the higher valency to precipitate as a brown hydrated oxide after addition of potassium iodide and sodium hydroxide. Upon acidification, manganese reverts to its divalent state and liberates iodine from potassium iodide, equivalent to the original dissolved oxygen content of the sample.

The liberated iodine is titrated against N/80 sodium thiosulphate using fresh iodine as an indicator.

Apparatus required: BOD bottles-300ml capacity, sampling devices, lab glassware - measuring cylinder, conical flasks, etc., and Bunsen burner.

Reagents:

- Manganese sulphate: 480g of manganous sulphate tetrahydrate is dissolved and made up to 1000ml with distilled water (Discarded if it changes colour with starch).
- Alkaline iodide-azide reagent: 500g of sodium hydroxide and 150g of potassium iodide along with 10g of sodium azide (NaN_3) is dissolved and made up to 1000ml with distilled water.
- Conc. sulphuric acid
- Starch indicator: 0.5g of starch is dissolved in distilled water and boiled for few minutes.

- Stock sodium thiosulphate: 24.82g of sodium thiosulphate pentahydrate ($Na_2S_20_2$. $5H_2O$) is dissolved in distilled water and made up to 1000ml.
- Standard sodium thiosulphate (0.025N): 250ml of the stock sodium thiosulphate pentahydrate is made up to 1000ml with distilled water to give 0.025N.

Procedure: The samples are collected in BOD bottles, to which 2ml of manganous sulphate and 2ml of potassium iodide are added and sealed. This is mixed well and the precipitate allowed to settle down. At this stage 2ml of conc. sulphuric acid is added, and mixed well until all the precipitate dissolves. 203ml of the sample is measured into the conical flask and titrated against 0.025N sodium thiosulphate using starch as an indicator. The end point is the change of colour from blue to colourless.

Calculations:

203ml because (200) (300)/ (200-4) = 203ml.

1ml of 0.025N Sodium thiosulphate = 0.2mg of Oxygen

$$\text{Dissolved} = \frac{(0.2)\ (1000\ \text{ml of Sodium thiosulphate})}{\text{Oxygen(as mg/L)}\ 200}$$

(Water analysis, APHA, 16th edn., pp 423-17)

Biological Oxygen Demand

Biological Oxygen Demand (BOD) is the amount of oxygen required by microorganisms for stabilizing biologically decomposable organic matter (carbonaceous) in water under aerobic conditions. The test is used to determine the pollution load of wastewater, the degree of pollution and the efficiency of wastewater treatment methods. 5-Day BOD test being a bioassay procedure (involving measurement of oxygen consumed by bacteria for degrading the organic matter under aerobic conditions) requires the addition of nutrients and maintaining the standard conditions of pH and temperature and absence of microbial growth inhibiting substances.

Principle: The method consists of filling the samples in airtight bottles of specified size and incubating them at specified temperature (20 °C) for 5 days. The difference in the dissolved oxygen measured initially and after incubation gives the BOD of the sample.

Apparatus required: BOD bottles - 300ml capacity, air incubator - to be controlled at 20 °C -\+ 1 °C, oximeter and magnetic stirrer.

Reagents:

- Preparation of dilution water: To 1000ml of water, 1ml each of phosphate buffer, magnesium sulphate, calcium chloride and ferric chloride solution is added, before bringing it to 20 °C and aerating it thoroughly.

Procedure: The sample having a pH of 7 is determined for first day DO. Various dilutions (at least 3) are prepared to obtain about 50% depletion of D.O. using sample and dilution water.

The samples are incubated at 20 °C for 5 days and the 5th day D.O is noted using the oximeter. A reagent blank is also prepared in a similar manner.

Calculation:

$$\text{BOD (in mg/L)} = \frac{(D_1 - D_2) - (B_1 - B_2) \times f}{p}$$

D_1 - 1st day D.O of diluted sample

D_2 - 5th day D.O of diluted sample

P - decimal volumetric fraction of sample used.

B_1 - 1st day D.O of control

B_2 - 5th day D.O of control

(Water analysis, APHA, 16th edn)

Chemical Oxygen Demand

Chemical oxygen demand (COD) is the measure of oxygen equivalent to the organic content of the sample that is susceptible to oxidation by a strong chemical oxidant. The intrinsic limitation of the test lies in its ability to differentiate between the biologically oxidisable and inert material. It is measured by the open reflux method.

Principle: The organic matter in the sample gets oxidized completely by strong oxidizing agents such as potassium dichromate in the presence of conc. sulphuric acid to produce carbon-di-oxide and water.

The excess potassium dichromate remaining after the reaction is titrated with Ferrous Ammonium Sulphate (FAS) using ferroin indicator to determine the COD. The dichromate consumed gives the oxygen required for the oxidation of the organic matter.

Apparatus required: Reflux apparatus, Nessler's tube, Erlenmeyer flasks, hot plate and lab glassware.

Reagents:

- Standard potassium dichromate solution (0.250M): 12.25g of potassium dichromate dried at 103 °C for about 2 hours is dissolved in distilled water and made up to 1000ml.
- Standard ferrous ammonium sulphate (FAS) 0.25N: 98g of FAS is dissolved in minimum distilled water to which 20ml of conc. sulphuric acid is added and made up to 1000ml using distilled water to give 0.25N of ferrous ammonium sulphate.
- Ferroin indicator: 1.485g of 1,10-Phenanthroline monohydrate and 695mg of ferrous sulphate is dissolved in 100ml of distilled water.
- Conc. sulphuric acid
- Silver sulphate crystals
- Mercuric sulphate crystals

Procedure: 15ml of conc. sulphuric acid with 0.3g of mercuric sulphate and a pinch of silver sulphate along with 5ml of 0.025M potassium dichromate is taken into a Nessler's tube. 10ml of sample (thoroughly shaken) is pipetted out into this mixture and kept for about 90 minutes on the hot plate for digestion. 40ml of distilled water is added to the cooled mixture (to make up to 50ml) and titrated against 0.25M FAS using ferroin indicator, till the colour turns from blue green to wine red indicating the end point. A reagent blank is also carried out using 10ml of distilled water.

Calculation:

$$\text{COD (mg/L)} = \frac{(\text{Blank reading - Sample reading}) \text{ X N X F X } 1000}{\text{Sample taken, ml}}$$

To calculate F,

$$F = \frac{10000}{\text{Titrant value of blank.}}$$

Fluoride

Fluorides have dual significance in water supplies. High concentration causes dental fluorosis and lower concentration (<0.8 mg/L) causes dental caries. A fluoride concentration of approximately 1mg/L in drinking water is recommended. They are frequently found in certain industrial processes resulting in fluoride rich wastewaters. Significant sources of fluoride are found in coke, glass and ceramic, electronics, pesticide and fertilizer manufacturing, steel and aluminium processing and electroplating industries. It is calculated by SPADNS method.

Principle: The colourimetric method of estimating fluoride is based on the reaction of fluorides (HF) with zirconium SPADNS solution and the 'lake' (colour of SPADNS reagent), which is greatly influenced by the acidity of the reaction mixture.

Fluoride reacts with the dye 'lake', dissociating (bleaching) the dye into a colourless complex anion (ZrF_6^{2-}). As the amount of fluoride increases, the colour produced becomes progressively higher or of different hue.

Apparatus required: Spectrophotometer and lab glassware.

Reagents:

- Standard fluoride solution.

  Stock: 221.0mg of AR grade sodium fluoride was dissolved in distilled water and made up to 1000ml to give 1ml = 100 m g of F^-
- *Working Standard:* 100ml of the stock fluoride was diluted to 1000ml to give 1ml = 10 mg of flouride.
- *SPADNS Solution:* 958mg of SPADNS is dissolved in 500ml of distilled water.1
- *Zirconyl-acid reagent:* 133mg Zirconyl chloride octahydrate ($ZrOCl_2 .8H_2O$) was dissolved in about 25ml of distilled water. 350ml of conc. HCl was added and diluted to 500ml with distilled water.
- *Zirconyl acid-SPADNS reagent:* Equal volume of SPADNS and zirconyl acid reagent was mixed.

Procedure: A standard graph is prepared by using fluoride concentrations ranging from 0.005 mg/L to 0.150 mg/L at 570nm. A reference solution is prepared by adding 4ml of acid zirconyl-SPADNS reagent to 21ml of distilled water. A known volume of filtered sample (21ml) is taken in a test tube, 4ml of acid zirconyl-SPADNS reagent is added to the sample along with a reference solution. The mixture is left for about 30 min for complete colour development and the optical density is read at 570nm.

Calculation:

$$F^- \text{ mg/L} = \frac{\text{(O.D sample) (Conc. of the Standard) (1000)}}{\text{(O.D Standard) (sample taken)}}$$

Free Carbon-di-oxide

The important source of free carbon-di-oxide in surface water bodies is mainly from respiration and decomposition by aquatic organisms. It reacts with water partly to form calcium bicarbonate and in the absence of bicarbonates gets converted to carbonates releasing carbon-di-oxide.

Principle: Free carbon-di-oxide reacts with sodium carbonate or sodium hydroxide to form sodium bicarbonate. The completion of the reaction is indicated by the development of pink colour, characteristic of phenolpthalein indicator at an equivalent pH of 8.3

Apparatus required: Lab glassware - measuring jar, pipette, conical flask etc.

Reagents:

- *Sodium hydroxide solution (0.22N):* 1g of sodium hydroxide was dissolved in 100ml of distilled water and made up to 1000ml to give 0.22N.
- *Phenolpthalein indicator*

Procedure: A known volume (50ml) of the sample is measured into a conical flask. 2-3 drops of phenolpthalein indicator is added and titrated against 0.22N sodium hydroxide till the pink colour persists indicating the end point.

Calculation:

$$\text{Free } CO_2 \text{ (mg/L)} = \frac{(Vt)(1000)}{Vs}$$

Where, Vt - volume of titrant (ml)

Vs - volume of the sample taken (ml)

Potassium

Potassium ranks seventh among the elements in order of abundance, behaves similar to sodium and remains low. Though found in small quantities (<20mg/L) it plays a vital role in the metabolism.

Principle: Trace amount of potassium can be determined by direct reading of flame photometer at a specific wavelength of 766.5nm by spraying the sample into the flame. The desired spectral lines are then isolated by the use of interference filters or suitable slit arrangements. The intensity of light is measured by the phototube.

Working principle of Flame photometer: The emission of characteristic radiations by alkali and alkaline earth metals and the correlation of the emission intensity with the concentration of the element form the basis of flame photometry. The principle of the flame photometer depends on the "Emission Spectroscopy" in which the electrons of the metals after absorbing energy get excited from ground state to higher energy level and return back to the ground state with emission of light. The sample under test is introduced into flame in solution by means of atomizer. The radiation from the flame enters a dispersing device and isolates it (radiation) from the flame to the desired region of the spectrum. The phototube measures the intensity of isolated radiation, which is proportional to the concentration of the element present in the sample.

Apparatus required: Flame photometer, lab glassware and Whattman filter paper.

Reagents:

- *Deionised distilled water.*
- *Stock potassium solution:* 1.907g of dried Potassium chloride, is dissolved in 1000ml of distilled water, to give 1ml = 1mg of potassium
- *Working Potassium solution*: Working standards of suitable strengths are prepared from the stock solution.

Procedure: The filter of the flame photometer is set at 766.5nm (marked for Potassium, K) the flame is adjusted for blue colour. The scale is set to zero and maximum using the highest standard value. A standard curve of different concentration is prepared by feeding the standard solutions. The sample is filtered through the filter paper and fed into the flame photometer. The concentration is found from the standard curve or as direct reading.

Sodium

Sodium is one of the most abundant elements and is a common constituent of natural waters. The sodium concentration of water is of concern primarily when considering their solubility for agricultural uses or boiler feed water. The concentration ranges from very low in the surface waters and relatively high in deep ground waters and highest in the marine waters. It is calculated by flame photometric method.

Reagents:

- *Deionised water*
- *Stock sodium solution:* 2.542g of Sodium chloride of dried (at 140° C) was dissolved in 1000ml distilled water to give 1ml = 1mg of sodium.

- *Working Potassium solution*: Working standards of suitable strengths were prepared from the stock solution.

Procedure: The filter of the flame photometer is set to 589nm (marked for Sodium, Na). By feeding distilled water the scale is set to zero and maximum using the standard of highest value. A standard curve between concentration and emission is prepared by feeding the standard solutions. The sample is filtered through filter paper and fed into the flame photometer and the concentration is found from graph or by direct readings.

Heavy Metals

Heavy metals are elements (properties of metals satisfied) of high atomic numbers. They have high utilities in industrial applications from papers to automobiles, by their very characteristic properties. They are found in the deep bowels of the earth as ores (complexes of mixtures). The metals are segregated from these ores, leaving behind the tailings that find their way into the environment as toxic pollutants. They get into the water bodies directly from point sources as sewage, and non-point sources as runoff and more insidiously as atmospheric deposition that are transported from long distances. Heavy metals affect every level of the food web, from producers in the trophic levels to the highest order carnivore by residing in the system and magnifying at every trophic status.

Atomic Absorption Spectrophotometer (AAS)

Working principle: Atomic absorption spectrometer's working principle is based on the sample being aspirated into the flame and atomized when the AAS's light beam is directed through the flame into the monochromator, and onto the detector that measures the amount of light absorbed by the atomized element in the flame. Since metals have their own characteristic absorption wavelength, a source lamp composed of that element is used, making the method relatively free from spectral or radiational interferences. The amount of energy of the characteristic wavelength absorbed in the flame is proportional to the concentration of the element in the sample.

Procedure: The sample is thoroughly mixed by shaking, and 100ml of it is transferred into a glass beaker of 250ml volume, to which 5ml of conc. nitric acid is added and heated to boil till the volume is reduced to about 15-20ml, by adding conc. nitric acid in increments of 5ml till all the residue is completely dissolved. The mixture is cooled, transferred and made up to 100ml using metal free distilled water.

Operations for analysing heavy metals:

- Lamp selection:
- Lamp for the element to be detected is selected.
- Operating current is suitably adjusted.
- The lamp is aligned for the visible beam to fall on the slit of the monochromator
- Wave length selection and slit adjustment:
- Appropriate wavelength for the element to be detected is selected. The wavelength controller is moved clockwise or anti-clockwise slowly to get maximum percentage of transmittance.
- The slits are adjusted to get closest to the required wavelength and avoid excess stray light.
- Flame adjustment:
- On selecting suitable wavelength, the acetylene-air mixture is lit at the recommended pressure.
- The burner level is so adjusted that the beam from the cathode crosses 1cm from the top of the burner and the beam is stabilized.
- Analysis:
- A calibration graph is obtained by feeding the standard solutions of suitable concentration.
- The samples are aspirated by feeding them through the capillary and the readings are noted. (Distilled water is aspirated between samples).

Apparatus required: AAS and lab glassware.

Reagents:

- Air-Acetylene flame
- Metal free water
- Standard metal solutions

Lead

Lead is relatively a minor element in the earth's crust but is widely distributed in low concentrations in uncontaminated soils and rocks. Lead concentration in freshwater is generally much higher. High concentration of lead results from atmospheric input of lead originating from its use in the leaded gasoline or from smelting processes.

Industrial processes such as printing and dyeing, paint manufacturing, explosives, photography and mine or smelter operations may contain relatively high values in lead. Lead is toxic to aquatic organisms.

Principle: Lead can be determined at a wavelength of 283.3 nm by AAS with aspiration of the sample into the oxidising air-acetylene flame. When the aqueous sample is aspirated, the sensitivity for 1% absorption is 0.5 mg/L and the detection limit is 0.05 mg/L.

Standard lead solution: 1.598g of lead nitrate is dissolved in about 200ml of water containing 1.5ml of conc. nitric acid and diluted to 1000ml of metal free water to give 1ml = 1mg lead.

A series of standards ranging from 1mg to 5mg were prepared from the stock and a standard graph was made.

Copper

Copper is a widely distributed trace element because most copper minerals are relatively insoluble and is sorbed to solid phases, hence only low concentrations are normally present in natural waters. Because of the presence of sulphide, copper would be expected to be even less soluble in anoxic systems.

The presence of higher concentrations of copper can usually be attributed to corrosion of copper pipes, industrial wastes or particularly in reservoirs, which uses copper as algaecides.

Copper is an essential trace element in the nutrition of plants and animals including man. It is required for the function of several enzymes and is necessary in the biosynthesis of chlorophyll. High levels are toxic to organisms but the response varies greatly with species.

Principle: Copper can be determined at a wavelength at 324.7 nm by AAS with aspiration of the sample into an oxidising air-acetylene flame. When the aqueous sample is aspirated, the sensitivity for 1% absorption is 0.1 mg/L and the detection limit is 0.01 mg/L.

Standard copper solution: 1g of copper salt is dissolved in 15ml of 1+1 nitric acid and diluted to 1000ml to give 1ml = 1mg copper. A series of standards ranging from 1mg to 5mg are prepared from the stock and a standard graph prepared.

Nickel

Standard nickel solution: 1.273g of nickel oxide is dissolved in a minimum quantity of 10% HCl and diluted to 1000ml with distilled water to give 1ml = 1mg of nickel. A series of standards ranging from 1mg to 5mg are prepared from the stock and analyzed.

The write up seems inadequate??

Iron

Iron is an abundant element in the earth's crust, but exists generally in minor concentrations in natural water systems. Iron is found in the +2 *(ferrous)* and +3 *(ferric)* states depending on the oxidation-reduction potentials of the water. The ferric state of iron imparts orange strain to any settling surfaces, including laundry articles, cooking and eating utensils, and plumbing fixtures.

Principle: Iron can be determined at a wavelength of 248.3 nm by AAS with aspiration of the sample into an oxidising air-acetylene flame. Under standard conditions, iron produces 1% absorption at 0.12 mg/L and a linear response up to about 5 mg/L.

Standard iron solution: 1g of iron is dissolved in 50ml of 1+1 nitric acid and diluted to 1000ml with distilled water to give 1ml = 1mg of iron. A series of standards ranging from 1mg to 5mg are prepared from the stock and analysed.

Chromium

The concentration of chromium in natural waters is usually very low. Elevated concentrations of chromium can result from mining and industrial processes. Chromate compounds are routinely used in cooling waters to control erosion. Chromium in water supplies is generally found in the hexavalent form.

Principle: Total chromium can be determined at a wavelength of 357.9 nm by atomic absorption with aspiration of sample into a reducing air-acetylene flame. Under standard concentrations, chromium produces 1 % absorption at 0.25 mg/L and is detectable down to 0.003 mg/L.

Standard chromium solution: 2.828g of AR grade potassium dichromate is dissolved in about 200ml of distilled water, with 1.5ml of conc. nitric acid and made up to 1000ml with the same to give 1ml = 1mg of chromium. A series of standards ranging from 1mg to 5mg are prepared from the stock and analysed.

Cadmium

Cadmium is largely found in nature in the form of sulphide, and as an impurity of zinc - lead ores. The abundance of cadmium is much less than that of zinc. Cadmium may enter the surface waters as a consequence of mining, electroplating plants, pigment works, textile and chemical industries, and is toxic to man. There is evidence that cadmium affects reproductive organs in humans and is also a potential carcinogen. A specific disease called "itai-itai" has been absorbed in

Japan due to excess cadmium. In addition, due to bioaccumulation, certain edible organisms may become hazardous to the ultimate consumer.

Principle: Cadmium can be determined at a wavelength of 228.8 nm by atomic absorption with aspiration of sample into an oxidising air-acetylene flame. When the aqueous sample is aspirated, the sensitivity for 1 % absorption is 25 mg/L and the detection limit is 2 mg/L.

Standard cadmium solution: 1.000 g of cadmium metal is dissolved in minimum volume of 1+1 HCl and made up to 1000ml with distilled water to give 1ml = 1mg of cadmium. A series of standards ranging from 1mg to 5mg are prepared from the stock and analysed.

Zinc

Zinc is an abundant element in rocks and ores and is present in natural waters only as a minor constituent. The main industrial use of zinc is in galvanizing and may enter the drinking waters from galvanized pipes. Another important use is in the preparation of alloys, including brass and bronze. It is an essential element in human nutrition. Food provides the main source of zinc to the body. Zinc may be toxic to aquatic organisms but the degree of toxicity varies greatly depending on water quality characteristics as well as the species being considered.

Principle: Zinc can be determined at a wavelength of 213.9 nm by AAS aspiration of the sample into an oxidizing air-acetylene flame. When the aqueous sample is aspirated, the sensitivity for 1% absorption is 20 mg/L and the detection limit is 5 mg/L.

Standard zinc solution: 1.000g of zinc metal is dissolved in 20ml of 1+1 HCl and diluted to 1000ml in distilled water, to give 1ml = 1mg of zinc. A series of standards ranging from 1mg to 5mg are prepared from the stock and analysed.

Sl. No	*Element*	*Wavelength (in nm)*
1	Lead	283.3
2	Copper	324.7
3	Iron	248.3
4	Chromium	357.9
5	Cadmium	228.8
6	Zinc	213.9
7	Nickel	??

Biological Parameters

Plankton Analysis

The physical and chemical characteristics of water affect the abundance, species composition, stability and productivity of the indigenous populations of aquatic organisms. The biological methods used for assessing water quality includes collection, counting and identification of aquatic organisms; biomass measurements; measurements of metabolic activity rates; toxicity tests; bioaccumulation; biomagnification of pollutants; and processing and interpretation of biological data. The work involving plankton analysis would help in:

1. Explaining the cause of colour and turbidity and the presence of objectionable odour, tastes and visible particles in waters.
2. The interpretation of chemical analyses.
3. Identifying the nature, extent and biological effects of pollution.
4. Providing data on the status of an aquatic system on a regular basis.

Plankton: A microscopic community of plants (phytoplankton) and animals (zooplankton), found usually free floating, swimming with little or no resistance to water currents, suspended in water, nonmotile or insufficiently motile to overcome transport by currents, are called "Plankton".

Phytoplankton (microscopic algae) usually occurs as unicellular, colonial or filamentous forms and is mostly photosynthetic and is grazed upon by the zooplankton and other organisms occurring in the same environment. Zooplankton principally comprise of microscopic protozoans, rotifers, cladocerans and copepods. The species assemblage of zooplankton also may be useful in assessing water quality.

The structure of photosynthetic populations in the aquatic ecosystems is dynamic and constantly changing in species composition and biomass distribution. An understanding of the community structure is dependent on the ability to understand the temporal distribution of the different species. Changes in species composition and biomass may affect photosynthetic rates, assimilation efficiencies, rates of nutrient utilization, grazing, etc.

Plankton, particularly phytoplankton, has long been used as indicators of water quality. Because of their short life spans, planktons respond quickly to environmental changes. They flourish both in highly eutrophic waters while a few others are very sensitive to organic and/

or chemical wastes. Some species have also been associated with noxious blooms causing toxic conditions apart from the tastes and odour problems.

Plankton net: The plankton net is a field-equipment used to trap plankton. It has a polyethylene filter of a defined mesh size and a graduated measuring jar attached to the other end. A handle holds the net. The mesh size of the net determines the size range of the plankton trapped. The mesh number 30 of size 60 mm was used for collecting samples.

Sampling Procedure: The manner in which sampling is done should conform to the objectives of the study. The "surface samples" (samples collected from the surface) are collected as close to the water surface as possible, mostly towards the centre of the lake at regular monthly intervals. A known volume of the sample, 5L to 50 L is filtered and planktons are filtered and preserved for further analysis

Labels: The sample label has the date, time of sampling, study area-lake name and the volume measured and pasted on the containers of 50ml capacity.

Preservation: The samples collected into the 100ml polyethylene vials were preserved by adding suitable amounts of 1ml chloroform to act as the narcotizing agent and 2ml of 2% formalin for preservation and analyses.

Concentration technique: The plankton nets are used to collect samples for the qualitative and quantitative estimation of the plankton, by filtering a known volume of water (5-50 liters) through the net depending on the plankton density of the tanks.

Qualitative and quantitative evaluation of plankton: Detailed analyses of phytoplanktonic populations are done by estimating the numbers in each species. The phytoplankton consisting of individual cells, filaments and colonies are counted as individual cells. When colonies of species are counted, the average number of cells per colony is counted, and in filamentous algae, the average length of the filament has to be determined.

Sedimentation and enumeration by microscope: Preserved samples in bottles are mixed uniformly by gentle inversion and then exactly 1ml of the sample is pipetted out into the S-C cell for analysis.

Microscope

Compound microscope: A monocular compound microscope is used in the counting of plankton with different eyepieces such as

10X, 15X and 20X. The microscope is calibrated using plankton-counting squares.

Counting

Counting cell- Sedgwick-Rafter (S-R) cell: The Sedgwick-Rafter cell is a devise used for plankton counting and is about 50mm long by 20mm wide and 1mm deep. The cell is covered by a relatively thick cover slip and is calibrated to contain exactly 1.0 ml.

Method

Filling the cell: The cover slip is placed diagonally across the S-R cell and filled with the sample carefully without air bubbles with a large bore pipette. The sample is allowed to settle for about 5 minutes before the actual counting begins.

Note: Since the configuration of the S-R cell does not allow the use of high power microscope objectives, the identification of organisms smaller than 10 – 15 mm is difficult or impossible, limiting the usage to only larger forms of relatively dense populations.

Strip counting: A "strip" is the length of the cell that constitutes a volume approximately 50 mm long, 1-mm deep accounting to the volume of 25mm^3 or 1/40 (2.5%) of the total cell volume.

By moving the mechanical stage from left to right, the organisms can be examined in a systematic manner.

By knowing the surface area of the portion counted in relation to that of the total, a factor is determined to expand the average counts of the plankton to the total area of the counting surface. This total area represents the number of organisms present per given volume of the sample. This volume expanded to an appropriate factor yields the organisms per litre of water for the lake.

The total number of planktons in the S-R cell is obtained by multiplying actual count in the strip by the number (enumeration factor) representing the portion of the S-R cell counted.

The number of the strips counted is a function of the precision desired and the number of units (cells, colonies) for the strips measured. In this study, 500 cells were counted for estimation.

The plankton count in the S-R cell is got from the following,

$$\text{Number/ml} = \frac{C \times 1000 \text{ mm}}{L \times D \times W \times S}$$

Where

C = Number of organisms counted

L = Length of each strip (S-R cell length) mm

D = Depth of a strip (S-R cell depth) mm

W = Width of a strip in mm

S = Number of strips counted

V_1 = (50)(1)(W)

= mm^3

The plankton counts per strip are then determined by multiplying the actual count by the factor representing the counted portion of the whole S-R cell volume.

Number/ml = (C) (1000 mm^3) / (L) (D) (W) (S)

Where

C = Number of organisms counted.

L = Length of each strip in mm (of S-R cell)

D = Depth of the strip in mm (S-R cell)

W = Width of the strip in mm (Whipple grid image width)

S = Number of strips counted.

Phytoplankton Counting Units: Some plankton are unicellular while others are multicellular (colonial), posing a problem for enumeration. For analysis, a colony of plankton is accounted as a single count. The large forms that cross two or more boundaries of the grid are counted separately at lower magnification and their number included in the total count.

Hypoxia (Environmental)

Hypoxia refers to low oxygen conditions. Normally 20.9% of the gas in the atmosphere is oxygen. The partial pressure of oxygen in the atmosphere is 20.9% of the total barometric pressure. In water however; oxygen levels are much lower, approximately 1%, and fluctuate locally depending on the presence of photosynthetic organisms and relative distance to the surface (there is more oxygen in the air, which will diffuse across the partial pressure gradient).

Atmospheric hypoxia occurs naturally at high altitudes. Total atmospheric pressure decreases as altitude increases, causing a lower partial pressure of oxygen which is defined as hypobaric hypoxia. Oxygen remains at 20.9% of the total gas mixture, differing from hypoxic hypoxia, where the percentage of oxygen in the air (or blood) is decreased. This is common, for example, in the sealed burrows of some subterranean animals, such as blesmols. Atmospheric hypoxica is also the basis of altitude training which is a standard part of training for elite athletes. Several companies mimic hypoxica using normobaric artificial atmosphere.

Aquatic Hypoxia

Oxygen depletion, is a phenomenon that occurs in aquatic environments as dissolved oxygen (DO; molecular oxygen dissolved in the water) becomes reduced in concentration to a point where it becomes detrimental to aquatic organisms living in the system. Dissolved oxygen is typically expressed as a percentage of the oxygen that would dissolve in the water at the prevailing temperature and salinity (both of which affect the solubility of oxygen in water).

An aquatic system lacking dissolved oxygen (0% saturation) is termed anaerobic, reducing, or anoxic; a system with low concentration—in the range between 1 and 30% saturation—is called hypoxic or dysoxic. Most fish cannot live below 30% saturation. A "healthy" aquatic environment should seldom experience less than 80%. The exaerobic zone is found at the boundary of anoxic and hypoxic zones.

Where Hypoxia Occurs

Hypoxia can occur throughout the water column and also at high altitudes as well as near sediments on the bottom. It usually extends throughout 20-50% of the water column, but depending on the water depth and location of pycnoclines (rapid changes in water density with depth). It can occur in 10-80% of the water column. For example, in a 10-meter water column, it can reach up to 2 meters below the surface. In a 20-meter water column, it can extend up to 8 meters below the surface.

Causes of Hypoxia

Oxygen depletion can result from a number of natural factors, but is most often a concern as a consequence of pollution and eutrophication in which plant nutrients enter a river, lake, or ocean, and phytoplankton blooms are encouraged. While phytoplankton, through photosynthesis, will raise DO saturation during daylight hours, the dense population of a bloom reduces DO saturation during the night by respiration. When phytoplankton cells die, they sink towards the bottom and are decomposed by bacteria, a process that further reduces DO in the water column. If oxygen depletion progresses to hypoxia, fish kills can occur and invertebrates like worms and clams on the bottom may be killed as well.

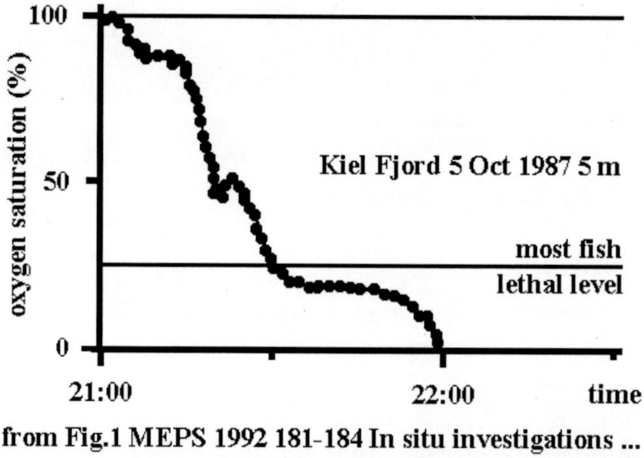

Figure: *Decline of oxygen saturation to anoxia, measured during the night in Kiel Fjord, Germany. Depth = 5 m*

Hypoxia may also occur in the absence of pollutants. In estuaries, for example, because freshwater flowing from a river into the sea is less dense than salt water, stratification in the water column can result. Vertical mixing between the water bodies is therefore reduced, restricting the supply of oxygen from the surface waters to the more saline bottom waters.

Figure: *Still frame from an underwater video of the sea floor. The floor is covered with crabs, fish, and clams apparently dead or dying from oxygen depletion.*

The oxygen concentration in the bottom layer may then become low enough for hypoxia to occur. Areas particularly prone to this include shallow waters of semi-enclosed water bodies such as the Waddenzee or the Gulf of Mexico, where land run-off is substantial. In these areas a so-called "dead zone" can be created. Low dissolved oyxgen conditions are often seasonal, as is the case in Hood Canal and areas of Puget Sound, in Washington State.

The World Resources Institute has identified 375 hypoxic coastal zones around the world, concentrated in coastal areas in Western Europe, the Eastern and Southern coasts of the US, and East Asia, particularly in Japan.

Hypoxia may also be the explanation for periodic phenomena such as the Mobile Bay jubilee, where aquatic life suddenly rushes to the shallows, perhaps trying to escape oxygen-depleted water. Recent widespread shellfish kills near the coasts of Oregon and Washington are also blamed on cyclic dead zone ecology.

Figure: *Jubilee photo from Mobile Bay*

Solutions

To combat hypoxia, it is essential to reduce the amount of land-derived nutrients reaching rivers in runoff. This can be done by improving sewage treatment and by reducing the amount of fertilizers leaching into the rivers. Alternately, this can be done by restoring natural environments along a river; marshes are particularly effective in reducing the amount of phosphorus and nitrogen (nutrients) in water. Other natural habitat-based solutions include restoration of shellfish populations, such as oysters. Oyster reefs remove nitrogen from the water column and filter out suspended solids, subsequently reducing the likelihood or extent of harmful algal blooms or anoxic conditions. Foundational work toward the idea of improving marine water quality through shellfish cultivation was conducted by Odd Lindahl et al., using mussels in Sweden.

Technological solutions are also possible, such as that used in the redeveloped Salford Docks area of the Manchester Ship Canal in England, where years of runoff from sewers and roads had accumulated in the slow running waters. In 2001 a compressed air injection system was introduced, which raised the oxygen levels in the water by up to 300%. The resulting improvement in water quality led to an increase in the number of invertebrate species, such as freshwater shrimp, to more than 30. Spawning and growth rates of fish species such as roach and perch also increased to such an extent that they are now amongst the highest in England.

In a very short time the oxygen saturation can drop to zero when offshore blowing winds drive surface water out and anoxic depthwater rises up. At the same time a decline in temperature and a rise in salinity is observed (from the longterm ecological observatory in the seas at Kiel Fjord, Germany). New approaches of long-term monitoring of oxygen regime in the ocean observe online the behaviour of fish and zooplankton, which changes drastically under reduced oxygen saturations (ecoSCOPE) and already at very low levels of water pollution.

Chapter 5

FishBash and Common Fishes Found in India

FishBase is a global species database of fish species (specifically finfish). It is the largest and most extensively accessed online database on adult finfish on the web. Over time it has "evolved into a dynamic and versatile ecological tool", widely cited in scholarly publications.

FishBase provides comprehensive species data, including information on taxonomy, geographical distribution, biometrics and morphology, behaviour and habitats, ecology and population dynamics as well as reproductive, metabolic and genetic data. There is access to tools such as trophic pyramids, identification keys, biogeographical modelling and fishery statistics and there are direct species level links to information in other databases such as LarvalBase, GenBank, the IUCN Red List and the Catalogue of Fishes.

As of August 2011, FishBase included descriptions of 32,100 species and subspecies, 292,500 common names in almost 300 languages, 50,500 pictures, and references to 46,000 works in the scientific literature. The site receives about 30 million hits per month

Albuliformes

The bonefishes are a family (Albulidae) of ray-finned fish that are popular as game fish in Florida, select locations in the South Pacific, and the Bahamas (where two bonefish are featured on the 10-cent coin) and elsewhere. The family is small, with 12 species in two genera.

Presently, the bonefishes are in their own order: Albuliformes. The spiny eels (Notacanthidae) and halosaurs (Halosauridae) were previously classified in this order, but are now, according to FishBase, given their own order, Notacanthiformes. The largest bonefish caught in the Western Hemisphere is a 16 pound, 3 ounce example caught off Islamorada, Florida, on March 19, 2007.

Description

The bonefishes' closest relatives are the tarpons and ladyfishes in the order Elopiformes. Bonefishes are unlike tarpons in that their mouths are under the snout rather than the end of it, and bonefishes lack the tarpons' protruding snout. Like tarpons and ladyfishes, bonefishes can breathe air via a modified swim bladder, and are found in brackish waters. Bonefish larvae are leptocephalic.

The slender body of the bonefish is silver, with a blue to green tinted back. On the upper half of the body there are dark streaks with cross bands connecting to the lateral line. Also, the body of a bonefish is rounded with a long downward aiming snout. The dorsal and caudal fins are black. Bonefish vary in adult length from 40–100 cm depending on species. The average size of a bonefish is from 3 to 5 pounds (1–2 kg) with the Florida record being 16 pounds 3 oz (7.34 kg).

The bonefishes are brackish or saltwater fish typically living in estuaries and travelling out to sea to spawn on a lunar cycle. They feed in the shallow sand and mud flats, on benthic organisms, such as worms, mollusks, shrimp, and crabs. They use their conical-shaped snouts to root out their prey, and can often be seen with their tails out of the water. Bonefishes possess crushing teeth in their palates.

Anguilliformes

Eels are an order of fish which consists of four suborders, 20 families, 111 genera and about 800 species. Most eels are predators. The term "eel" (originally referring to the European eel) is also used for some other similarly shaped fish, such as electric eels and spiny eels, but these are not members of the Anguilliformes order.

Eels are elongated fish, ranging in length from 5 cm (2.0 in) in the one-jawed eel (*Monognathus ahlstromi*) to 4 m (13 ft) in the slender giant moray. Adults range in weight from 30 g (1.1 oz) to well over 25 kg (55 lb). They possess no pelvic fins, and many species also lack pectoral fins. The dorsal and anal fins are fused with the caudal fin, forming a single ribbon running along much of the length of the animal.

Most eels live in the shallow waters of the ocean and burrow into sand, mud, or amongst rocks. A majority of eel species are nocturnal, thus are rarely seen. Sometimes, they are seen living together in holes, or "eel pits". Some species of eels also live in deeper water on the continental shelves and over the slopes deep as 4,000 m (13,000 ft). Only members of the *Anguilla* regularly inhabit fresh water, but they, too, return to the sea to breed.

Eels begin life as flat and transparent larvae, or leptocephali. Eel larvae drift in the surface waters of the sea, feeding on marine snow, small particles that float in the water. Eel larvae then metamorphose into glass eels and then become elvers before finally seeking out their juvenile and adult habitats. Freshwater elvers travel upstream and are forced to climb up obstructions, such as weirs, dam walls, and natural waterfalls. Lady Colin Campbell found, at Ballisodare, the eel fisheries were greatly improved by the hanging of loosely plaited grass ladders over barriers, enabling the elvers to ascend.

The heaviest true eel is the European conger. The maximum size of this species has been reported as reaching a length of 3 m (10 ft) and a weight of 110 kg (240 lb). Other eels are longer, but do not weigh as much, such as the slender giant moray which reaches 4 m (13 ft).

Anguillidae (Freshwater eels)

The Anguillidae are a family of fishes that contains the freshwater eels. The 19 species and six subspecies in this family are all in genus *Anguilla*. They are catadromous, meaning they spend their lives in freshwater rivers, lakes, or estuaries, and return to the ocean to spawn. The young eel larvae, called leptocephali, live only in the ocean and consume small particles called marine snow. They grow larger in size, and in their next growth stage, they are called glass eels. At this stage, they enter estuaries, and when they become pigmented, they are known as elvers. Elvers travel upstream in freshwater rivers, where they grow to adulthood. Some details of eel reproduction are as yet unknown, and the discovery of the spawning area of the American and European eels in the Sargasso Sea is one of the more famous anecdotes in the history of ichthyology. The spawning areas of some other anguillid eels, such as the Japanese eel, and the giant mottled eel, were also discovered recently in the western North Pacific Ocean.

Freshwater eels are elongated with tubelike, snake-shaped bodies. They have large, pointed heads and their dorsal fins are usually continuous with their caudal and anal fins, to form a fringe lining the posterior end of their bodies. They have small pectoral fins to help them navigate along river bottoms. Their scales are thin and soft. Freshwater eels go through physical changes in their bodies when going to and from the ocean for different stages of life.

Anguillid eels are important food fish. Eel aquaculture is a fast-growing industry. Important food eel species include longfin eel, Australian long-finned eel, short-finned eel, and Japanese eel. Most eel production historically has been in Japan, Korea, and Taiwan, but in recent years, the greatest production has been in China.

Seafood Watch, one of the most well-known sustainable seafood advisory lists, recommends consumers avoid eating anguillid eels due to significant pressures on worldwide populations. Several species used as *unagi* have seen their population sizes greatly reduced in the past half century. Catches of the European eel, for example, have declined about 80% since the 1960s. Although about 90% of freshwater eels consumed in the US are farm-raised, they are not bred in captivity. Instead, young eels are collected from the wild and then raised in various enclosures. In addition to wild eel populations being reduced by this process, eels are often farmed in open net pens which allow parasites, waste products, and diseases to flow directly back into wild eel habitat, further threatening wild populations. Freshwater eels are carnivores and as such are fed other wild-caught fish, adding another element of unsustainability to current eel-farming practices.

Anguilla Bengalensis Bengalensis

The Indian mottled eel, *Anguilla bengalensis bengalensis*, is a subspecies of eel in the genus *Anguilla* of the family Anguillidae. It is found throughout the Indian subcontinent and neighbouring regions including the East Indies.

Showing the typical characteristics of the Anguillidae, this species grows to 1.2 m and as much as 6 kg. Dorsal fin soft rays number 250–305, anal fin soft rays 220–250, vertebrae between 106 and 112 in number.

The Indian mottled eel is valued as a food fish. The mucus of this eel is used in a medicine for arthritis. It is known by numerous common names in the native languages of the regions it inhabits.

Colocongridae

The Colocongridae, the worm eels or short-tail eels are a family of eels, containing a single genus, *Coloconger*.

Colongrids are found in tropical waters of the Atlantic, Indian, and West Pacific oceans. They are bottom-dwelling fish, living in waters from 300 to 900 m (980 to 2,950 ft) in depth. Compared with other eels, they have relatively short and stubby bodies, with blunt snouts.

Congridae (Conger and Garden eels)

The Congridae are the family of conger and garden eels. Congers are valuable and often large food fishes, while garden eels live in colonies, all protruding from the sea floor after the manner of plants in a garden (thus the name). The family includes over 180 species in 32 genera.

The European conger, *Conger conger*, is the largest of the family; it has been recorded at up to 3 m (9.8 ft) in length and weighing 350 lb (160 kg).

Congrids are found in tropical, subtropical and temperate seas across the world. Clear distinguishing features among congrids are few; they all lack scales, and most possess pectoral fins. They feed on crustaceans and small fish, and unlike some other eels, do not migrate to breed.

- *Muraenesocidae (Pike congers)*
- *Muraenidae (Moray eels)*
- *Nemichthyidae (Snipe eels)*
- *Nettastomatidae (Duckbill eels)*
- *Ophichthidae (Snake eels)*

Atheriniformes

The Atheriniformes, also known as the silversides, are an order of ray-finned fishes that includes the Old World silversides and several less-familiar families, including the unusual Phallostethidae. They are found worldwide in tropical and temperate marine and freshwater environments. Atheriniformes are generally elongated and silvery in colour, although exceptions do exist. They are typically small fish, with the largest being the jacksmelt, with a head-body length of 44 cm (17 in), and the smallest species, such as the Bangkok minnow, being only 2 cm (0.79 in) in adult length.

Members of the order usually have two dorsal fins, the first with flexible spines, and an anal fin with one spine at the front. The lateral line is typically weak or absent.

Atheriniform larvae share several characteristics; the gut is unusually short, a single row of melanophores occurs along the back, and the fin rays do not become evident until some time after hatching. They scatter their eggs widely, with most species attaching them to aquatic plants.

Classification of the atheriniformes is uncertain, with the best evidence for monophyly in the larval characteristics mentioned below. Their closest relatives are thought to be the Cyprinodontiformes.

Following Nelson (2006), the family Melanotaeniidae includes the subfamilies Bedotiinae, Melanotaeniinae, Pseudomugilinae, and Telmatherininae, to demonstrate their monophyly.

However, in a 2004 study, a different classification scheme classifies the families Bedotiidae, Melanotaeniidae, and Pseudomugilidae (also include Telmatherinine genera) in a suborder Melanotaenioidei. Thus, the number of families in Atheriniformes varies from author to author.

Atherinidae (Silversides)

The Old World silversides are a family, Atherinidae, of fish in the order Atheriniformes. They occur worldwide in tropical and temperate waters. About two-thirds of the species are marine, and the remainder live in fresh water. The 71 species are in 14 genera. The genus *Craterocephalus* is the most diverse with 25 species. Five genera are monotypic.

Silversides are relatively small, with the largest species, the jacksmelt, *Atherinopsis californiensis*, reaching 44 cm (17 in), while most are under 20 cm (7.9 in), and several are not recorded at more than 5 cm (2.0 in).

The body is generally elongated. Distinctive characters include two dorsal fins widely separated, with the first consisting of flexible spines and the second having one spine followed by soft rays, while the anal fin has one spine on the leading edge followed by soft rays. The pectoral fins tend to be high, and there is no lateral line. On the flanks is a broad, silvery band. The scales are relatively large.

They feed on zooplankton. Some species, such as the hardyhead silverside, *Atherinomorus lacunosus*, are commercially fished.

The family Atherinopsidae (Neotropical silversides) is closely related.

Cypriniformes

Cypriniformes is an order of ray-finned fish, including the carps, minnows, loaches and relatives. This order contains five to six families, over 320 genera, and more than 3,250 species, with new species being described every few months or so, and new genera being recognized frequently. They are most diverse in southeastern Asia, but are entirely absent from Australia and South America.

Their closest living relatives are the Characiformes (characins and allies), the Gymnotiformes (electric eel and American knifefishes) and the Siluriformes (catfishes).

Description

Like other orders of the Ostariophysi, fishes of cypriniformes possess a Weberian apparatus. However, they differ from most of their relatives in having only a dorsal fin on their back; most other fishes of Ostariophysi have a small fleshy adipose fin behind the dorsal fin. Further differences are the Cypriniformes' kinethmoid and the lack of teeth in the mouth. Instead, they have convergent structures called pharyngeal teeth in the throat.

While other groups of fish, such as cichlids, also possess pharyngeal teeth, the cypriniformes' teeth grind against a chewing pad on the base of the skull, instead of an upper pharyngeal jaw.

The most notable family placed here is Cyprinidae (carps and minnows) which make up two-thirds of the order's diversity. This is one of the largest families of fish, and is widely distributed across Africa, Eurasia, and North America. Most species are strictly freshwater inhabitants, but a considerable number are found in brackish water, such as roach and bream.

At least one species is found in saltwater, the Pacific redfin, *Tribolodon brandtii*. Brackish water and marine cyprinids are invariably anadromous, swimming upstream into rivers to spawn. Sometimes separated as family Psilorhynchidae, they seem to be specially-adapted fishes of Cyprinidae.

Balitoridae and Gyrinocheilidae are families of mountain stream fishes feeding on algae and small invertebrates. They are found only in tropical and subtropical Asia. While the former are a speciose group, the latter contain only a handful of species. The suckers (Catostomidae) are found in temperate North America and eastern Asia. These large fishes are similar to carps in appearance and ecology.

Members of Cobitidae common across Eurasia and parts of North Africa. A mid-sized group like the suckers, they are rather similar to catfish in appearance and behaviour, feeding primarily off the substrate and equipped with barbels to help them locate food at night or in murky conditions. Fishes in the familes Cobitidae, Balitoridae, Botiidae, and Gyrinocheilidae are called loaches, although it seems that the last do not belong to the lineage of "true" loaches but are related to the suckers.

Systematics

Historically these included all the forms now placed in the superorder Ostariophysi except the catfish, which were placed in the order Siluriformes.

By this definition, the Cypriniformes were paraphyletic, so recently the orders Gonorhynchiformes, Characiformes (characins and allies), and Gymnotiformes (knifefishes and electric eels) have been separated out to form their own monophyletic orders.

Figure: Nemacheilus chrysolaimos *is a stone loach. Closely related to true loaches, like these they have barbels.*

Figure: *The Chinese algae eater (*Gyrinocheilus aymonieri*), one of the sucking loaches which are distant from other "loaches"*

Figure: Erimyzon sucetta, *a small sucker*

The families of Cypriniformes are traditionally divided into two superfamilies. Superfamily Cyprinioidea contains the carps and minnows (Cyprinidae) and, according to some, also the mountain carps as the family Psilorhynchidae.

In 2012 Maurice Kottelat reviewed the superfamily Cobitioidea and under his revision it now consists of the following families: hillstream loaches (Balitoridae), Barbuccidae, Botiidae, suckers (Catostomidae), true loaches (Cobitidae), Ellopostomatidae, Gastromyzontidae, sucking loaches (Gyrinocheilidae), stone loaches (Nemacheilidae), Serpenticobitidae and long-finned loaches (Vaillantellidae).

Catostomoidea is usually treated as a junior synonym of Cobitioidea. But it seems that it could be split off the Catostomidae and Gyrinocheilidae in a distinct superfamily; the Catostomoidea might be closer relatives of the carps and minnows than of the "true" loaches. While the Cyprinioidea seem more "primitive" than the loach-like forms, they were apparently successful enough never to shift from the original ecological niche of the basal Ostariophysi.

Yet, from the ecomorphologically conservative main lineage apparently at least two major radiations branched off. These diversified from the lowlands into torrential river habitats, acquiring similar habitus and adaptations in the process.

The mountain carps are highly apomorphic Cyprinidae, perhaps close to true carps (Cyprininae), or maybe to the danionins. While some details about the phylogenetic structures of this massively diverse family are known – e.g. that Cultrinae and Leuciscinae are rather close relatives and stand apart from Cyprininae – there is no good consensus yet on how the main lineages are interrelated. A systematic list, from the most ancient to the most modern lineages, can thus be given as:

Superfamily Cobitioidea

- Family Balitoridae – hillstream loaches
- Family Cobitidae – true loaches

Superfamily Catostomoidea

- Family Catostomidae – suckers
- Family Gyrinocheilidae – sucking loaches

Superfamily Cyprinioidea

- Family Cyprinidae – carps (including koi and goldfish) and minnows (including Psilorhynchidae)

Evolution

Cypriniformes include the most primitive of the Ostariophysi in the narrow sense (i.e. excluding Gonorynchiformes). This is evidenced not only by physiological details, but their great distribution, which indicates they had the longest time to spread. The earliest that Cypriniformes might have diverged from Characiphysi (Characiformes and relatives) is thought to be about the Early Triassic, about 250 million years ago (mya). However, their divergence probably occurred only with the splitting-up of Pangaea in the Jurassic, maybe 160 million years ago. By 110 mya, the plate tectonics evidence indicates that the Laurasian Cypriniformes must have been distinct from their Gondwanan relatives.

Cypriniformes is thought to have originated in south-east Asia, where the most diversity of this group is found today. The alternative hypothesis is that they began in South America, similar to the other otophysans. If this were the case, they would have spread to Asia through Africa or North America before the continents split up, for these are purely freshwater fishes. As the Characiformes began to diversify and spread, they may have out-competed South American basal cypriniforms in Africa, where more advanced cypriniforms survive and coexist with characiforms.

The earliest cypriniform fossils are already assignable to the living family Catostomidae; from the Paleocene of Alberta, they are roughly 60 million years old. During the Eocene (55-35 mya), catostomids and cyprinids spread throughout Asia. In the Oligocene, around 30 mya, advanced cyprinids began to out-compete catostomids wherever they were sympatric, causing a decline of the suckers. Cyprinids reached North America and Europe by about the same time, and Africa in the early Miocene (some 23-20 mya). The cypriniforms spread to North America through the Bering land bridge, which formed and disappeared again several times during the many millions of years of cypriniform evolution.

Relationship with Humans

The Cyprinidae in particular are important in a variety of ways. Many species are important food fish, particularly in Europe and Asia. Some are also important as aquarium fish, of which the goldfish and koi are perhaps the most celebrated. The other families are of less commercial importance. The Catostomidae have some importance in angling, and some "loaches" are bred for the international aquarium fish trade.

Accidentally or deliberately introduced populations of common carp (*Cyprinus carpio*) and grass carp (*Ctenopharyngodon idella*) are found on all continents except Antarctica. In some cases, these exotic species have a negative impact on the environment. Carp in particular stir up the riverbed reducing the clarity of the water, making it difficult for plants to grow.

In science, one of the most famous members of the Cypriniformes is the zebrafish (*Danio rerio*). The zebrafish is one of the most important vertebrate model organisms in biological and biochemical sciences, being used in many kinds of experiment. As, during early development, the zebrafish has a nearly transparent body, it is ideal for studying developmental biology. It is also used for the elucidation of biochemical signalling pathways. They are also good pets, but can be shy in bright light and crowded tanks.

Threats and Extinction

Figure: *The thicktail chub (Gila crassicauda) is globally extinct since about 1960.*

Habitat destruction, damming of upland rivers, pollution and in some cases overfishing for food or the pet trade have driven some Cypriniformes to the brink of extinction or even beyond. In particular, Cyprinidae of southwestern North America have been severely affected; a considerable number went entirely extinct after settlement by Europeans. For example, in 1900 the thicktail chub (*Gila crassicauda*) was the most common freshwater fish found in California; 70 years later not a single living individual existed.

The well-known red-tailed black shark (*Epalzeorhynchos bicolor*) from the Mae Klong river of *The Bridge on the River Kwai* fame possibly only survives in captivity. Ironically, while pollution and other forms of overuse by humans have driven it from its native home, it is bred for the aquarium fish trade by the thousands. The Yarqon bleak (*Acanthobrama telavivensis*) from the Yarqon River had to be rescued into captivity from imminent extinction; new populations have apparently been established again successfully from captive stock. Balitoridae and Cobitidae, meanwhile, contain a very large number of species about which essentially nothing is known except how they look and where they were first found.

Balitoridae (River Loaches)

The hillstream loaches or river loaches are a family, the Balitoridae, of small fish from South, Southeast and East Asia. They are sometimes sold as "lizardfish" or (in Germany) "flossensaugers". Many of the species are popular for aquaria. They have a number of similarities with the Cobitidae, their sibling family of "loaches", such as multiple barbels around the mouth. They should not be confused with the loricariids, which look similar but are a family of catfish.

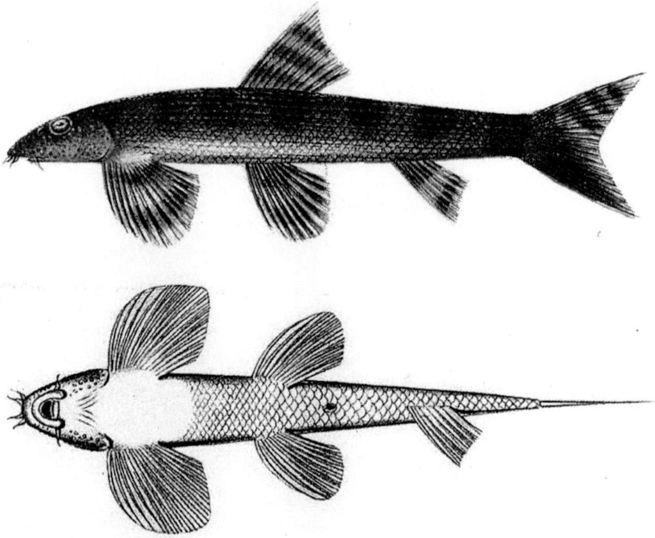

Most species are rheophilic, living in swift, clear and well-oxygenated streams. Several species of the family live in fast-flowing streams or torrents and have modified ventral fins used for clinging to rocks.

The subfamily Nemacheilinae has recently been separated as a distinct family, Nemacheilidae (stone loaches) and several genera have been separated into the family Gastromyzontidae.

Cyprinidae

Cyprinidae is a large family of freshwater fishes, including the carps, the true minnows, and their relatives (for example, the barbs and barbels). Commonly called the carp family or the minnow family, its members are also known as cyprinids. It is the largest fish family and the largest family of vertebrate animals in general, with over 2,400 species in about 220 genera. The family belongs to the order

Cypriniformes, of whose genera and species the cyprinids make up two-thirds. The family name is derived from the Ancient Greek *kyprînos* ("carp").

Figure: *Giant barbs (*Catlocarpio siamensis*) are the largest members of this family.*

Cyprinids are stomachless fish with toothless jaws. Even so, food can be effectively chewed by the gill rakers of the specialized last gill bow. These pharyngeal teeth allow the fish to make chewing motions against a chewing plate formed by a bony process of the skull. The pharyngeal teeth are species-specific and are used by specialists to determine species. Strong pharyngeal teeth allow fish such as the common carp and ide to eat hard baits like snails and bivalves.

Hearing is a well-developed sense, since the cyprinids have the Weberian organ, three specialized vertebral processes that transfer motion of the gas bladder to the inner ear. This construction is also used to observe motion of the gas bladder due to atmospheric conditions or depth changes. The cyprinids are physostomes because the pneumatic duct is retained in adult stages and the fish are able to gulp air to fill the gas bladder or they can dispose excess gas to the gut.

The fish in this family are native to North America, Africa, and Eurasia. The largest cyprinid in this family is the giant barb (*Catlocarpio siamensis*), which may grow up to 3 m (9.8 ft). The largest North American species is the Colorado pikeminnow (*Ptychocheilus lucius*), of which individuals up to 6 ft (1.8 m) long and weighing over 100 lb (45 kg) have been recorded.

Conversely, many species are smaller than 5 cm (2.0 in). As of 2008, the smallest known freshwater fish is a cypriniform, *Danionella translucida*, reaching 12 mm (0.47 in) at the longest. All fish in this family are egg-layers and most do not guard their eggs; however, a few species build nests and/or guard the eggs. The bitterling-like cyprinids (Acheilognathinae) are notable for depositing their eggs in bivalve molluscs, where the young grow until able to fend for themselves.

Most cyprinids feed mainly on invertebrates and vegetation, probably due to the lack of teeth and stomach, but some species, like the asp, specialize in fish. Many species (ide, common rudd) will eat small fish when they reach a certain size. Even small species, such as the moderlieschen, eat larvae of the common frog in artificial circumstances.

Some fish, such as the grass carp, are specialized in eating vegetation; others, such as the common nase, eat algae from hard surfaces, while others, such as the black carp, specialize in snails, and some, such as the silver carp, are specialized filter feeders. For this reason, they are often introduced as a management tool to control various factors in the aquatic environment, such as aquatic vegetation and diseases transmitted by snails.

Relationship with Humans

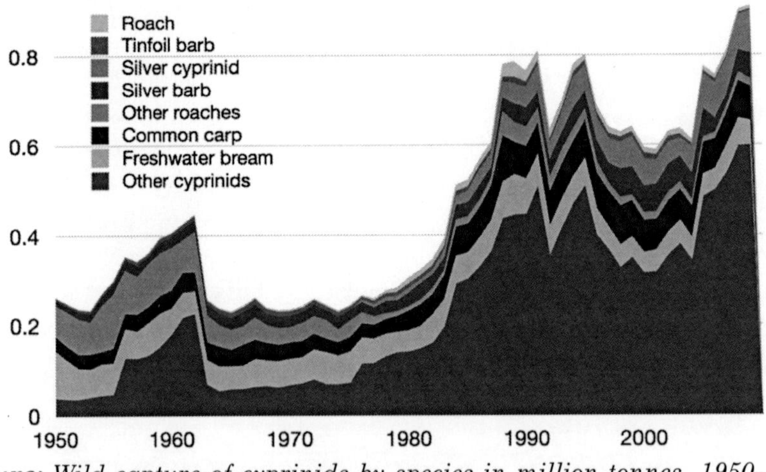

Figure: *Wild capture of cyprinids by species in million tonnes, 1950–2009, as reported by the FAO*

Cyprinids are highly important food fish; they are fished and farmed across Eurasia. In land-locked countries in particular, cyprinids are often the major species of fish eaten because they make the largest part of biomass in most water types except for fast-flowing rivers. In Eastern Europe, they are often prepared with traditional methods, such as drying and salting. The prevalence of inexpensive frozen fish products made this less important now than it was in earlier times.

Nonetheless, in certain places, they remain popular for food, as well as recreational fishing, and have been deliberately stocked in ponds and lakes for centuries for this reason.

Cyprinids are popular for angling especially for match fishing (due to their dominance in biomass and numbers) and fishing for common carp because of its size and strength.

Several cyprinids have been introduced to waters outside their natural ranges to provide food, sport, or biological control for some pest species. The common carp (*Cyprinus carpio*) and the grass carp (*Ctenopharyngodon idella*) are the most important of these, for example in Florida. In some cases, they have become invasive species that compete with native fishes or disrupt the environment. Carp in particular can stir up sediment, reducing the clarity of the water and making it difficult for plants to grow.

Numerous cyprinids have become important in the aquarium hobby, most famously the goldfish, which was bred in China from the Prussian carp (*Carassius (auratus) gibelio*). First imported into Europe around 1728, it was much fancied by Chinese nobility as early as 1150 AD and after it arrived there in 1502, also in Japan.

In the latter country, from the 18th century onwards, the common carp was bred into the ornamental variety known as koi – or more accurately *nishikigoi* (&"É›), as *koi* (É›) simply means "common carp" in Japanese.

Other popular aquarium cyprinids include danionins, rasborines, and true barbs. Larger species are bred by the thousands in outdoor ponds, particularly in Southeast Asia, and trade in these aquarium fishes is of considerable commercial importance. The small rasborines and danionines are perhaps only rivalled by characids and poecilid livebearers in their popularity for community aquaria.

One particular species of these small and undemanding danionines is the zebrafish (*Danio rerio*). It has become the standard model species for studying developmental genetics of vertebrates, in particular fish.

Habitat destruction and other causes have reduced the wild stocks of several cyprinids to dangerously low levels; some are already entirely extinct. In particular, the Leuciscinae from southwestern North America have been hit hard by pollution and unsustainable water use in the early to mid-20th century; most globally extinct Cypriniformes species are in fact Leuciscinae from the southwestern United States and northern Mexico.

Systematics

Figure: *Rainbow shark,* Epalzeorhynchos frenatum, *a semiagressive aquarium fish*

The massive diversity of cyprinids has so far made it difficult to resolve their phylogeny in sufficient detail to make assignment to subfamilies more than tentative in many cases. Some distinct lineages obviously exist – for example, Cultrinae and Leuciscinae, regardless of their exact delimitation, are rather close relatives and stand apart from Cyprininae –, but the overall systematics and taxonomy of the Cyprinidae remain a subject of considerable debate. A large number of genera are *incertae sedis*, too equivocal in their traits and/or too little-studied to permit assignment to a particular subfamily with any certainty.

Part of the solution seems that the delicate rasborines are the core group, consisting of minor lineages that have not shifted far from their evolutionary niche, or have coevolved for millions of years. These are among the most basal lineages of living cyprinids. Other "rasborines" are apparently distributed across the diverse lineages of the family.

The validity and circumscription of proposed subfamilies like Labeoninae or Squaliobarbinae also remains doubtful, although the latter do appear to correspond to a distinct lineage. The sometimes-seen grouping of the large-headed carps (Hypophthalmichthyinae) with *Xenocypris*, though, seems quite in error. More likely, the latter are part of the Cultrinae.

The entirely paraphyletic "Barbinae" and the disputed Labeoninae might be better treated as part of the Cyprininae, forming a close-knit group whose internal relationships are still little known. The small African "barbs" do not belong in *Barbus sensu stricto* – indeed, they are as distant from the typical barbels and the typical carps (*Cyprinus*) as these are from *Garra* (which is placed in the Labeoninae by most who accept the latter as distinct) and thus might form another as yet unnamed subfamily. However, as noted above, how various minor lineages tie into this has not yet been resolved; therefore, such a radical move, though reasonable, is probably premature.

Figure: *The tench (Tinca tinca) is of unclear affiliations and often placed in a subfamily of its own.*

The tench (*Tinca tinca*), a significant food species farmed in western Eurasia in large numbers, is unusual. It is most often grouped with the Leuciscinae, but even when these were rather loosely circumscribed, it always stood apart. A cladistic analysis of DNA sequence data of the S7 ribosomal protein intron 1 supports the view that it is distinct enough to constitute a monotypic subfamily. It also suggests it may be closer to the small East Asian *Aphyocypris*, *Hemigrammocypris*, and *Yaoshanicus*. They would have diverged roughly at the same time from cyprinids of east-central Asia, perhaps as a result of the Alpide orogeny that vastly changed the topography of that region in the late Paleogene, when their divergence presumably occurred.

A DNA-based analysis of these fish places Rasborinae as the basal lineage with Cyprininae as a sister clade to Leuciscinae. The subfamilies Acheilognathinae, Gobioninae, and Leuciscinae are monophylitic.

Cyprinodontiformes

The Cyprinodontiformes are an order of ray-finned fish, comprising mostly small, freshwater fish. Many popular aquarium fish, such as killifish and live-bearers, are included. They are closely related to the Atheriniformes and are occasionally included with them. A colloquial term for the order as a whole is toothcarps, though they are not actually close relatives of the true carps – the latter belong to the superorder Ostariophysi, while the toothcarps are Acanthopterygii.

The families of Cyprinodontiformes can be divided into three groups: viviparous and ovoviviparous (all species give live birth), and oviparous (all species egg-laying). The live-bearing groups differ in whether the young are carried to term within (ovoviviparous) or without (viviparous) an enclosing eggshell. Phylogenetically however, one of the two suborders – the Aplocheiloidei – contains oviparous species exclusively, as do two of the four superfamilies of the other suborder (the Cyprinodontoidea and Valencioidea of the Cyprinodontoidei). Vivipary and ovovivipary have evolved independently from oviparous ancestors, the latter possibly twice.

Description

Figure: Fundulopanchax scheeli, *a killifish of the Aplocheilidae.*

Members of this order are notable for inhabiting harsh environments, such as saline or very warm waters, water of poor quality, or isolated situations where no other types of fish occur. They are typically omnivores, and often live near the surface, where the oxygen-rich water compensates for environmental disadvantages.

They are small to medium-sized fish, with small mouths, large eyes, a single dorsal fin, and a rounded caudal fin. The largest species is the *cuatro ojos* (*Anableps dowi*), which measures 34 cm (13 in) in length, while the smallest, the least killifish (*Heterandria formosa*), is just 8 mm (0.31 in) long as an adult.

Elopiformes

The Elopiformes are the order of ray-finned fish including the tarpons, tenpounders, and ladyfish, as well as a number of extinct types. They have a long fossil record, easily distinguished from other fishes by the presence of an additional set of bones in the throat.

They are related to the order of eels, although the adults superficially resemble very large or giant herrings in appearance. The larvae, however, are leptocephali, looking very similar to those of eels.

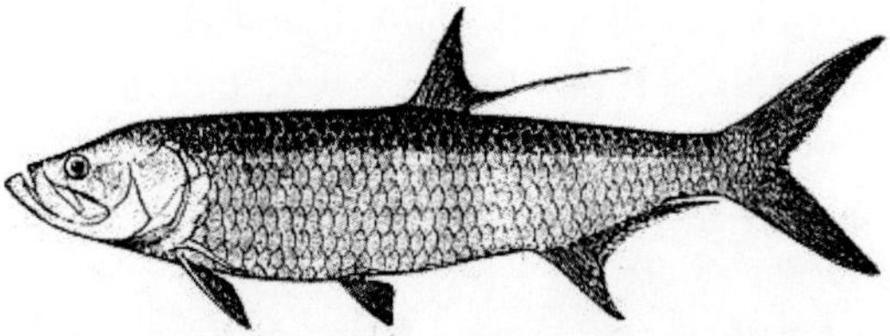

Figure: *Tarpon,* Megalops *sp.*

Although many fossil forms are known, the order is relatively small today, containing just two genera and nine species:

Order Elopiformes

- Family Elopidae (ladyfish)
  - o Genus *Elops* (seven species, one of which was described in 2010)
- Family Megalopidae (tarpons)
  - o Genus *Megalops* (two species)

Elopidae (Tenpounders)

The Elopidae are a family of ray-finned fish containing the single genus *Elops*. They are commonly known as ladyfish, skipjacks, jack-rashes, or tenpounders.

The ladyfish are a coastal-dwelling fish found throughout the tropical and subtropical regions, occasionally venturing into temperate waters. Spawning takes place at sea, and the fish larvae migrate inland entering brackish waters. Their food is smaller fish and crustaceans (shrimp). Typically throughout the species, the maximum size is 1 m (3.3 ft) and the maximum weight 10 kg (22 lb). The body is fusiform (tapering spindle shape) and oval in cross-section; being slightly laterally compressed, and the eyes are large and partially covered with adipose eyelids.

Like those of eels, the larvae are leptocephalic - being highly compressed, ribbon-like, and transparent. After initial growth, they shrink and then metamorphise into the adult form.

This family is fished, but the body is bony and therefore this fish is not marketed widely for consumption. They are caught and used as bait or may be ground down for fish meal.

Gadiformes

Gadiformes are an order of ray-finned fish, also called the Anacanthini, that includes the cod and its allies. Many major food fish are in this order. They are found in marine waters throughout the world and the vast majority of the species are found in temperate or colder regions (tropical species are typically deep-water). A few species may enter estuaries and a single, the burbot (*Lota lota*), is a freshwater fish.

Common characteristics include the positioning of the pelvic fins (if present), below or in front of the pectoral fins. Gadiformes are physoclists, which means their gas bladders do not have a pneumatic duct. The fins are spineless. Gadiform fish range in size from the codlets, which may be as small as 7 cm (2.8 in) in adult length, to the Atlantic cod, *Gadus morhua*, which reaches up to 2 m (6.6 ft).

Macrouridae (Grenadiers or Rattails)

Grenadiers or rattails (less commonly whiptails) are generally large, brown to black gadiform marine fish of the family Macrouridae. Found at great depths from the Arctic to Antarctic, members of this family were amongst the most abundant of the deep-sea fish.

The Macrouridae are a large and diverse family with some 34 genera and 383 species recognized (well over half of which are contained in just three genera, *Caelorinchus*, *Coryphaenoides*, and *Nezumia*). They range in length from about 10 cm (3.9 in) in the graceful grenadier (*Hymenocephalus gracilis*) to 1.5 m (4.9 ft) in the giant grenadier, *Albatrossia pectoralis*. An important commercial fishery exists for the larger species, such as the giant grenadier and roundnose grenadier, *Coryphaenoides rupestris*. The family as a whole may represent up to 15% of the deep-sea fish population.

Typified by large heads with large mouths and eyes, grenadiers have slender bodies that taper greatly to very thin caudal peduncles or tails (excluding one species with no tail fin): this rat-like tail explains the common name 'rattail' and the family name Macrouridae, from the Greek *makros* meaning "great" and *oura* meaning "tail". The first dorsal fin is small, high, and pointed (and may be spinous); the second dorsal fin runs along the rest of the back and merges with the tail and extensive anal fin. The scales are small.

As with many deep-living fish, the lateral line system in grenadiers is well-developed; it is further aided by numerous chemoreceptors located on the head and lips, and chemosensory barbels underneath the chin. Benthic species have gas bladders with unique muscles attached to them. The animals are thought to use these muscles to "strum" their gas bladders and produce sound, possibly playing a role in courtship and mate location. Light-producing organs, photophores, are present in some species; they are located in the middle of the abdomen, just before the anus and underneath the skin.

Living at depths from 200 to 6,000 m (660 to 19,690 ft), grenadiers are the most common benthic fish of the deep (however, two genera are known to prefer the midwater). They may be solitary or may form large schools, as with the roundnose grenadiers. The benthic species are attracted to structural oases, such as hydrothermal vents, cold seeps, and shipwrecks. They are thought to be generalists, feeding on smaller fish, pelagic crustaceans such as shrimp, amphipods, cumaceans and less often cephalopods and lanternfish. As well as being important apex predators in the benthic habitat, some species are also notable as scavengers.

As few rattail larvae have been recovered, little is known of their life histories. They are known to produce a large number (over 100,000) of tiny (1–2 millimetres or 0.039–0.079 inches in diameter) eggs made buoyant by lipid droplets. The eggs are presumed to float up to the thermocline (the interface between warmer surface waters and cold, deeper waters) where they develop. The juveniles remain in shallower waters, gradually migrating to greater depths with age.

Spawning may or may not be tied to the seasons, depending on the species. At least one species, *Coryphaenoides armatus*, is thought to be semelparous; that is, the adults die after spawning. Nonsemelparous species may live to 56 years or more. The Macrouridae in general are thought to have low resilience; commercially exploited species may be overfished and this could soon lead to a collapse of their fisheries.

Gasterosteiformes

Gasterosteiformes is an order of ray-finned fishes that includes the sticklebacks and relatives.

In the Gasterosteiformes, the pelvic girdle is never attached to the cleithra directly, and the supramaxillary, orbitosphenoid, and basisphenoid bones are absent. The body is often partly or completely covered with dermal plates.

The name "Gasterosteiformes" means "bone-bellies". It is derived from Ancient Greek *gaster*. The ending for fish orders "-formes" is derived from Latin and indicates "of similar form".

Often Gasterosteiformes includes the sea horses, pipefishes and their relatives as suborder Syngnathoidei, with the sticklebacks and relatives in the suborder Gasterosteoidei. In many more recent treatments the traditional placement as different orders, with the former becoming Syngnathiformes, is used. There is a growing body of evidence that this is correct; in fact, the two groups do not seem to be particularly close relatives among the Acanthopterygii.

As it seems, the loosely delimited Gasterosteiformes are paraphyletic with the Scorpaeniformes. The more typical members of that group (e.g. scorpionfishes) are apparently closer to the "true" Gasterosteiformes, whereas the keel-bodied flying gurnards (Dactylopteridae) seem actually to belong to the Syngnathiformes clade. It seems that the closest living relatives of the narrowly-delimited Gasterosteiformes are the gunnels (Pholidae) and eelpouts (Zoarcidae), traditionally placed in the massively paraphyletic "Perciformes". These two families, as well as the related Trichodontidae, would then appear to be derived offshoots of the scorpaeniform-gasterosteiform radiation which have apomorphically lost the bone "armour" found in their relatives.

Three families certainly belong to the Gasterosteiformes *sensu stricto*:

Aulorhynchidae – Tube-snouts

- Gasterosteidae – sticklebacks
- Hypoptychidae – the Sand Eel
- Indostomidae – armoured sticklebacks

The armoured sticklebacks (Indostomidae) and the dragonfishes and sea moths (Pegasidae) are variously placed with the pipefish or the stickleback lineage. While the Pegasidae are almost certainly Syngnathiformes, the placement of the monotypic former family is essentially unresolved, as it has always been.

Chanidae (Milkfish)

The milkfish (*Chanos chanos*) is the sole living species in the family Chanidae, in which about seven extinct species in five additional genera have been reported. The Hawaiian name for the fish is *awa* — not to be confused with *'awa*, the word for kava (*Piper methysticum*), which has an initial glottal stop. It is called *bangús* in the Philippines, where it is the national fish. In the Nauruan language, it is referred to as *ibiya*.

The milkfish has a generally symmetrical and streamlined appearance, with a sizable forked caudal fin. They can grow to 1.70 m (5 ft 7 in), but are most often about 1 m (39 in) in length. They have no teeth and generally feed on algae and invertebrates.

They occur in the Indian Ocean and across the Pacific Ocean, tending to school around coasts and islands with reefs. The young fry live at sea for two to three weeks and then migrate to mangrove swamps, estuaries, and sometimes lakes, and return to sea to mature sexually and reproduce.

Consumption

Figure: Bandeng duri lunak *(tender boned), Indonesian pressure cooked milkfish*

Figure: *Philippines grilled* bangus *(milkfish)*

The milkfish is an important seafood in Southeast Asia and some Pacific Islands. Because milkfish is notorious for being much bonier than other food fish, deboned milkfish, called "boneless *bangús*" in the Philippines, has become popular in stores and markets.

Another popular presentation of milkfish in Indonesia is *bandeng duri lunak* or *bandeng presto* (*ikan bandeng* is the Indonesian name for milkfish) from Central and East Java. *Bandeng presto* is pressure cooked milkfish until the bones are rendered tender. Another way to prepare milkfish is *bandeng asap* or smoked milkfish. Either fresh or processed, milkfish is the popular seafood product of Indonesian fishing towns, such as Juwana near Semarang in Central Java, and Sidoarjo near Surabaya in East Java.

Aquaculture

History

Milkfish aquaculture first occurred around 800 years ago in the Philippines and spread in Indonesia, Taiwan, and into the Pacific. Traditional milkfish aquaculture relied upon restocking ponds by collecting wild fry. This led to a wide range of variability in quality and quantity between seasons and regions.

In the late 1970s, farmers first successfully spawned breeding fish. However, they were hard to obtain and produced unreliable egg viability. In 1980, the first spontaneous spawning happened in sea cages. These eggs were found to be sufficient to generate a constant supply for farms.

Osteoglossiformes

Osteoglossiformes (Greek: "bony tongues") is a relatively primitive order of ray-finned fish that contains two sub-orders, the Osteoglossoidei and the Notopteroidei. All of the living species inhabit freshwater. They are found in South America, Africa, Australia and southern Asia, having first evolved in Gondwana before that continent broke up.

The mooneyes (Hiodontidae) are often classified here, but may also be placed in a separate order Hiodontiformes.

Members of the order are notable for having toothed or bony tongues, and for having the forward part of the gastrointestinal tract pass to the left of the oesophagus and stomach (for all other fish it passes to the right). In other respects, osteoglossiform fishes vary considerably in size and form; the smallest is *Pollimyrus castelnaui*, at just 2 centimetres (0.79 in) long, while the largest, the arapaima (*Arapaima gigas*), reaches as much as 2.5 metres (8.2 ft).

Notopteridae (Featherbacks or Knifefishes)

The family Notopteridae contains ten species of osteoglossiform (bony-tongued) fishes, commonly known as featherbacks and knifefishes. They are small fishes living in freshwater or brackish environments in Africa, and South and Southeast Asia.

With the denotation of "knifefish", the Notopterids should not be confused with Gymnotiforms, the electric knifefishes from South and

Central America. Although their manner of swimming is similar and they are superficially similar in appearance, the two groups are not closely related.

A few of the larger species, especially *Chitala ornata* , are food fish and occasionally aquarium pets. The name is from Greek *noton* meaning "back" and *pteron* meaning "fin".

Featherbacks have slender, elongated, bodies, giving them a knife-like appearance. The caudal fin is small and fused with the anal fin, which runs most of the length of the body. Where present, the dorsal fin is small and narrow, giving rise to the common name of "featherback". The fish swims by holding its body rigid and rippling the anal fin to propel itself forward or backwards.

Notopterids have specialized swim bladders. The organ extends throughout the body and even into the fins in some cases. Although the swim bladder is not highly vascularised, it can absorb oxygen from air and also functions to produce sound by squeezing air through a narrow passage into the pharynx.

Perciformes

Perciformes, also called the Percomorphi or Acanthopteri, are the largest order of vertebrates, containing about 41% of all bony fish. Perciformes means "perch-like". They belong to the class of ray-finned fish, and comprise over 10,000 species found in almost all aquatic ecosystems.

The order contains about 160 families, which is the most of any order within the vertebrates. It is also the most variably sized order of vertebrates, ranging from the 7-mm (1/4-in) *Schindleria brevipinguis* to the marlins in the *Makaira* genus. They first appeared and diversified in the Late Cretaceous.

Among the well-known members of this group are cichlids, California sheephead, bluegill, damselfish, bass, and perch.

The dorsal and anal fins are divided into anterior spiny and posterior soft-rayed portions, which may be partially or completely separated. The pelvic fins usually have one spine and up to five soft rays, positioned unusually far forward under the chin or under the belly. Scales are usually ctenoid, although sometimes they are cycloid or otherwise modified.

Classification is controversial. As traditionally defined, the Perciformes are almost certainly paraphyletic. Other orders that should possibly be included as suborders are the Scorpaeniformes,

Tetraodontiformes, and Pleuronectiformes. Of the presently recognized suborders, several may be paraphyletic, as well. These are grouped by suborder/superfamily, generally following the text *Fishes of the World.*

Murrel Fishes

These are air breathing fishes with long cylindrical body, flattened head and protractile mouth. These can grow in fresh water ponds, irrigation canals, wells and marshy areas. They breed even before the onset of monsoons.

Channa punctatus or Ophiocephalus:(Snake head) : It is a long fish with snake like body and accessory respiratory organs. As it lives outside the water also, it is commonly called Livefish. It is coloured differently. It grows to an average length of 30 - 35 cm. It is a common food fish of high demand.

Channa striatus: (stiped snake head): The body is coloured dark brown with yellow bands on either side. The fish feeds on worms and insects and grows to a length of 0.9 mt. Its flesh is good for health as it does not contain cholesterol.

Channa marulius: (large headed snake fish): It is also used for culture in fresh water ponds and tanks.

Exotic Fishes

When the indigenous fishnet are not favoured for culture due to economic viability, exotic breeds are selected and cultured. These fishes yield nutritious food and earn foreign exchange.

Cyprinus carpio: (Common carp) : This fish was imported from China and introduced into Nilagiri lakes. The growth of the fish is higher and grows to a length of 75cm and 6.5 kg weight. It breeds thrice in a year. When cultured under extensive system, the productivity was at 1500 kg/ha.

Osphronemus goramy: (Gowramy): It was imported from Jawa and Maritius and introduced into fresh waters of Madras and Calcutta. The rate of growth is very slow.

Ctenopharyngodon idella: (Grass carp): It grows in fresh water, polluted water and brackish water of low salinity. It feeds on aquatic weeds and used to eliminate them. It is native of Japan and China and was introduced into Cuttack waters in 1959. It can grow to a size of 0.9m in size and 7kg in weight.

Hypothalamychthys molitrix: (silver fish): It was imported from Hongkong and introduced into fresh water of Cuttack region. The mouth is located dorsally at the tip of snout, the body is laterally compressed and enclosed by small shiny scales. It feeds on the left over food particles of carp fish and grows quickly. It reaches to a length of 60 cm and weighs about 1.5 kg.

Tilapia mossambicus: It was imported from East Africa in 1952. The upper jaw in males is larger. It breeds even at the age of two months.

Cold Water Fishes

These fishes are commonly called sport fishes. These fishes include trouts and Mahseers.

Salmogiardneri: (Rain bow trout): It is a north American fish introduced into rivers and lakes of Ooty and bill of kerala. The dorsal and caudal fins are pinkish with dark spots. It grows to a length of 1 .8m. and weigh upto 100kg..

Tortor: (Mahseer): Head possess short rosiral and long maxillary barbs. Dorsal ride of the body is Mahseer-Tortor-fish grayish green, lateral sides are gold and belly is silvery white.

Even though it is adapted to grow well in canals but now it is cultured in reservoirs like Bakranangal. It grows to a size of 1 mt. and considered as good food fish.

Tor Khudree: Snout is pointed. Body dark coloured on dorsal and lateral sides while yellow on ventrolateral sides.

Tinca tinca (Doctor fish): It is the native of Europe and West Siberia and was introduced in Indian waters. It grows to a size of 40cm.

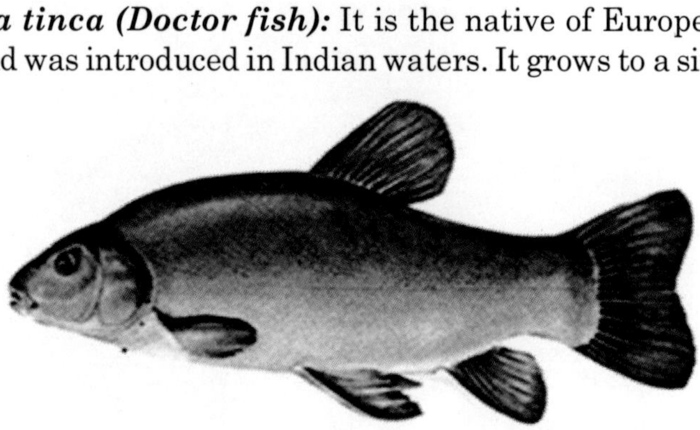

Description and Habitat

The fish is marine; freshwater; brackish; pelagic-neritic; anadromous ; depth range ? - 200 m. Within a tropical range; 34°N - 5°N, 42°E - 97°E in marine and freshwater. It can grow up to 60 cm in length with weights of up to 3 kg. It is found in rivers and estuaries in India, Pakistan, Bangladesh, Burma and the Persian Gulf area where it can be found in the Tigris and Euphrates rivers in and around Iran and Iraq. It has no dorsal spines but 18 - 21 Dorsal soft rays and anal soft rays. The belly has 30 to 33 scutes. There is a distinct median notch in upper jaw. Gill rakers fine and numerous, about 100 to 250 on lower part of arch and the fins are hyaline. The fish shows a dark blotch behind gill opening, followed by a series of small spots along the flank in juveniles. Colour in life, silver shot with gold and purple. The species filter feeds on plankton and by grubbing muddy bottoms. The fish schools in coastal waters and ascends up the rivers (anadromous) for around 50 – 100 km to spawn during the South West monsoons (June to September) and also in January to March . The young fish returning to the sea are known in Bangladesh as jatka, which includes any hilsa fish up to 9 inches long.

Food Value

The fish is popular food amongst the people of South Asia and in the Middle East, but especially with Bengalis, and it is the national fish of Bangladesh. Bengali fish curry is a popular dish made with mustard oil or seed. It is also popular in India, especially West Bengal, Odisha, Tripura, Assam Andhra Pradeshand Southern Gujarat and in Mizoram and it is also exported globally.

In North America (where ilish is not always readily available) other shad fish are sometimes used as an ilish substitute, especially in Bengali cuisine. This typically occurs near the East coast of North America, where fresh shad fish having similar taste can be found.

In Bangladesh, fish are caught in the Padma-Meghna-Jamuna delta, which flows into the Bay of Bengal and Meghna (lower Brahmaputra), and Jamuna rivers. In India, the Rupnarayan (which has the *Kolaghater Ilish*), Ganges, Mahanadi, Chilka Lake, Narmada and Godavari rivers are also famous. They are also caught in the sea, but some consider the marine stage of the fish as not so tasty. The fish has very sharp and tough bones, making it problematic to eat for some.

Ilish is an oily fish rich in omega 3 fatty acids. Recent experiments have shown its beneficial effects in decreasing cholesterol level in rats and insulin level.

In Bengal, ilish can be smoked, fried, steamed, baked in young plantain leaves, prepared with mustard seed paste, curd, Begun (eggplant), different condiments like jira and so on. It is said that people can cook ilish in more than 50 ways. Ilish roe is also popular as a side dish. Ilish can be cooked in very little oil since the fish itself is very oily.

Ilish in Culture

In Odisha there is a popular saying that "Machha khaaiba Ilishii, chakiri kariba polisi", which means that eating hilsa and getting a job in police department are of equal status, meaning that eating hilsais respectable. In Andhra Pradesh, the saying goes "Pustelu ammi ayina Pulasa tinocchu", meaning It's worth eating Pulasa/Hilsa by even selling the nuptials.

In pohela boisakh (the first day of the Bangali New Year), it is customary to have ilish with panta bhat (watery rice left overnignt) typically at breakfast. The meal is the traditional way to celebrate the Bangladeshi new year.

In many Bengali Hindu families a pair of ilish fishes (Bengali: *Joda Ilish*) are bought on auspicious days, for example for special prayers or pujadays like for the Hindu Goddess of music, art and knowledge Saraswati Puja, which takes place in the beginning of Spring or on the day of Lakshmi Puja (The Goddess of Wealth and Prosperity) which takes place in autumn. Some people offer the fish to the goddess Lakshmi, without which the Puja is sometimes thought to be incomplete.

Ilish Production

Five type of ilish can be found worldwide. Yearly ilish caught are 5,000,000 ton. Among them, 50%-60% are caught by Bangladesh, 15%-20% are caught by India, Pakistan and rest 5%-10% are caught by Malaysia, Thailand, China, Vietnam and Sri Lanka.

Catfish

Catfishes (order Siluriformes) are a diverse group of ray-finned fish. Named for their prominent barbels, which resemble a cat's whiskers, catfish range in size and behaviour from the heaviest and longest, the Mekong giant catfish from Southeast Asia and the second longest, the wels catfish of Eurasia, to detritivores (species that eat dead material on the bottom), and even to a tiny parasitic species commonly called the candiru, *Vandellia cirrhosa*.

There are armour-plated types and there are also naked types, neither having scales. Despite their name, not all catfish have prominent barbel. Members of the Siluriformes order are defined by features of the skull and swimbladder.

Catfish are of considerable commercial importance; many of the larger species are farmed or fished for food. Many of the smaller species, particularly the genus *Corydoras*, are important in the aquarium hobby. Many catfish are nocturnal, but others (many *Auchenipteridae*) are *crepuscular* or *diurnal* (most *Loricariidae or Callichthyidae* for example).

Distribution and Habitat

Extant catfish species live inland or in coastal waters of every continent except Antarctica. Catfish have inhabited all continents at one time or another. Catfish are most diverse in tropical South America, Asia and Africa with one family native to North America and one family in Europe. More than half of all catfish species live in the Americas. They are the only ostariophysans that have entered freshwater habitats in Madagascar, Australia, and New Guinea.

They are found in freshwater environments, though most inhabit shallow, running water. Representatives of at least eight families are hypogean (live underground) with three families that are also troglobitic (inhabiting caves). One such species is *Phreatobius cisternarum*, known to live underground in phreatic habitats. Numerous species from the families Ariidae and Plotosidae, and a few species from among the Aspredinidae and Bagridae, are found in salt water.

In the United States, catfish species may be known by a variety of slang names, such as "mud cat", "polliwogs", or "chuckleheads". These nicknames are not standardized, so one area may call a bullhead catfish by the nickname "chucklehead", while in another state or region, that nickname refers to the blue catfish.

Physical Characteristics

External anatomy of catfish: Most catfish are bottom feeders. In general, they are negatively buoyant, which means that they will usually sink rather than float due to a reduced gas bladder and a heavy, bony head. Catfish have a variety of body shapes, though most have a cylindrical body with a flattened ventrum to allow for benthic feeding.

A flattened head allows for digging through the substrate as well as perhaps serving as a hydrofoil. Some have a mouth that can expand to a large size and contains no incisiform teeth; catfish generally feed through suction or gulping rather than biting and cutting prey. However, some families, notably Loricariidae and Astroblepidae, have a suckermouth that allows them to fasten themselves to objects in fast-moving water. Catfish also have a maxilla reduced to a support for barbels; this means that they are unable to protrude their mouths as other fish such as carp.

Catfish may have up to four pairs of barbels: nasal, maxillary (on each side of mouth), and two pairs of chin barbels, even though pairs of barbels may be absent depending on the species. Many larger catfish also have chemoreceptors across their entire bodies, which means they "taste" anything they touch and "smell" any chemicals in the water.

"In catfish, gustation plays a primary role in the orientation and location of food". Because their barbels and chemoreception are more important in detecting food, the eyes on catfish are generally small. Like other ostariophysans, they are characterized by the presence of a Weberian apparatus. Their well-developed Weberian apparatus and reduced gas bladder allow for improved hearing as well as sound production.

Catfish have no scales; their bodies are often naked. In some species, the mucus-covered skin is used in cutaneous respiration, where the fish breathes through its skin. In some catfish, the skin is covered in bony plates called scutes; some form of body armor appears in various ways within the order. In loricarioids and in the Asian genus *Sisor*, the armor is primarily made up of one or more rows of free dermal plates. Similar plates are found in large specimens of *Lithodoras*. These plates may be supported by vertebral processes, as in scoloplacids and in *Sisor*, but the processes never fuse to the plates or form any external armor. By contrast, in the subfamily Doumeinae (family Amphiliidae) and in hoplomyzontines (Aspredinidae), the armor is formed solely by expanded vertebral processes that form plates. Finally, the lateral armor of doradids, *Sisor*, and hoplomyzontines consists of hypertrophied lateral line ossicles with dorsal and ventral lamina.

All catfish, except members of Malapteruridae (electric catfish), possess a strong, hollow, bonified leading spine-like ray on their dorsal and pectoral fins. As a defence, these spines may be locked into place so that they stick outwards, which can inflict severe wounds. In several species catfish can use these fin rays to deliver a stinging protein if the fish is irritated. This venom is produced by glandular cells in the epidermal tissue covering the spines. In members of the family Plotosidae, and of the genus *Heteropneustes*, this protein is so strong it may hospitalize humans, those unfortunate enough to receive a sting; in *Plotosus lineatus*, the stings may result in death.

Juvenile catfish, like most fish, have relatively large heads, eyes and posterior median fins in comparison to larger, more mature individuals. These juveniles can be readily placed in their families, particularly those with highly derived fin or body shapes; in some cases identification of the genus is possible. As far as known for most catfish, features that are often characteristic of species such as mouth and fin positions, fin shapes, and barbel lengths show little difference between juveniles and adults. For many species, pigmentation pattern is also similar in juveniles and adults. Thus, juvenile catfishes generally resemble and develop smoothly into their adult form without distinct juvenile specializations. Exceptions to this are the ariid catfishes, where the young retain yolk sacs late into juvenile stages, and many pimelodids, which may have elongated barbels and fin filaments or colouration patterns.

Sexual dimorphism is reported in about half of all families of catfish. The modification of the anal fin into an intromittent organ (in internal fertilizers) as well as accessory structures of the reproductive apparatus (in both internal and external fertilizers) have been described in species belonging to 11 different families.

Size: Catfish have one of the greatest ranges in size within a single order of bony fish. Many catfish have a maximum length of under 12 cm. Some of the smallest species of Aspredinidae and Trichomycteridae reach sexual maturity at only 1 centimetre (0.39 in).

Figure: *A sting from the striped eel catfish, Plotosus lineatus, may be fatal.*

The wels catfish, *Silurus glanis*, is the only native catfish species of Europe, besides the much smaller related Aristotle's catfish found in Greece. Mythology and literature record wels catfish of astounding proportions, yet to be proven scientifically. The average size of the species is about 1.2–1.6 m (3.9–5.2 ft), and fish more than 2 metres (6.6 ft) are very rare. The largest specimens on record measure more than 2.5 metres (8.2 ft) in length and sometimes exceeded 100 kilograms (220 lb). In July 2009, a catfish weighing 193 pounds (88 kg) was caught in the River Ebro, Spain, by an 11-year old British schoolgirl.

In North America the largest *Ictalurus furcatus* (Blue catfish) caught in the Missouri River on 20 July 2010, weighed 130 pounds (59 kg). The largest flathead catfish, *Pylodictis olivaris*, ever caught was in Independence, Kansas, weighing 123 lb 9 oz (56.0 kg).

These records pale in comparison to a giant Mekong catfish caught in northern Thailand on 1 May 2005 and reported to the press almost

2 months later that weighed 293 kilograms (646 lb). This is the largest giant Mekong catfish caught since Thai officials started keeping records in 1981. This species is not well studied since it lives in developing countries and it is quite possible it can grow even larger.

Also in Asia, Jeremy Wade caught a 75.5 kilograms (166.4 lb) Goonch following three fatal attacks on humans in the Kali River on the India-Nepal border. Wade was of the opinion that the offending fish must even have been significantly larger to have taken an 18 year old boy as well as a water buffalo.

Internal anatomy: In many catfish, the *humeral process* is a bony process extending backward from the pectoral girdle immediately above the base of the pectoral fin. It lies beneath the skin where its outline may be determined by dissecting the skin or probing with a needle.

The retina of catfish are composed of single cones and large rods. Many catfish have a tapetum lucidum which may help enhance photon capture and increase low-light sensitivity. Double cones, though present in most teleosts, are absent from catfish.

The anatomical organization of the testis in catfish is variable among the families of catfish, but the majority of them present fringed testis: Ictaluridae, Claridae, Auchenipteridae, Doradidae, Pimelodidae, and Pseudopimelodidae. In the testes of some species of Siluriformes, organs and structures such as a spermatogenic cranial region and a secretory caudal region are observed, in addition to the presence of seminal vesicles in the caudal region. The total number of fringes and their length are different in the caudal and cranial portions between species. Fringes of the caudal region may present tubules, in which the lumen is filled by secretion and spermatozoa. Spermatocysts are formed from cytoplasmic extensions of Sertoli cells; the release of spermatozoa is allowed by breaking of the cyst walls.

The occurrence of seminal vesicles, in spite of their interspecific variability in size, gross morphology and function, has not been related to the mode of fertilization. They are typically paired, multi-chambered, and connected with the sperm duct, and have been reported to play a glandular and a storage function. Seminal vesicle secretion may include steroids and steroid glucuronides, with hormonal and pheromonal functions, but it appears to be primarily constituted of mucoproteins, acid mucopolysaccharides, and phospholipids.

Fish ovaries may be of two types: gymnovarian or cystovarian. In the first type, the oocytes are released directly into the coelomic cavity and then eliminated. In the second type, the oocytes are conveyed to the exterior through the oviduct. Many catfish are cystovarian in type, including *Pseudoplatystoma corruscans*, *P. fasciatum*, *Lophiosilurus alexandri*, and *Loricaria lentiginosa*.

Communication

Sound production and interpretation: Catfish can produce different types of sounds and also have well-developed auditory reception used to discriminate between sounds with different pitches and velocities. They are also able to determine the distance of the sound's origin and from what direction it originated. This is a very important fish communication mechanism, especially during agonistic and distress behaviours. Catfish are able to produce a variety of sounds for communication that can be classified into two groups: drumming sounds and stridulation sounds. The variability in catfish sound signals differs due to a few factors: the mechanism by which the sound is produced, the function of the resulting sound, and physiological differences such as size, sex, and age. To create a drumming sound, catfish use an indirect vibration mechanism using a swimbladder. In these fishes, sonic muscles insert on the ramus Mulleri, also known as the elastic spring. The sonic muscles pull the elastic spring forward and extend the swimbladder. When the muscles relax, the tension in the spring quickly returns the swimbladder to its original position, which produces the sound.

Catfish also have a sound-generating mechanism in their pectoral fins. Many species in the catfish family possess an enhanced first pectoral fin ray, called the spine, which can be moved by large abductor and adductor muscles. The base of the catfishes' spines has a sequence of ridges, and the spine normally slides within a groove on the fish's pelvic girdle during routine movement; but, pressing the ridges on the spine against the pelvic girdle groove creates a series of short pulses. The movement is analogous to a finger moving down the teeth of a comb, and consequently a series of sharp taps is produced.

Sound-generating mechanisms are often different between genders. In some catfishes, pectoral fins are longer in males than in females of similar length, and differences in the characteristic of the sounds produced were also observed. Comparison between families of the same order of catfish demonstrated family and species-specific patterns of vocalization, according to a study by Maria Clara Amorim. During courtship behaviour in three species of *Corydoras* catfishes, all males actively produced stridulation sounds before egg fertilization, and the species' songs were different in pulse number and sound duration.

Sound production in catfish may also be correlated with fighting and alarm calls. According to a study by Kaatz, sounds for disturbance (e.g. alarm) and agonistic behaviour were not significantly different, which suggests distress sounds can be used to sample variation in agonistic sound production.

However, in a comparison of a few different species of tropical catfish, some fish put under distress conditions produced a higher intensity of stridulatory sounds than drumming sounds. Differences in the proportion of drumming versus stridulation sounds depend on morphological constraints, such as different sizes of drumming muscles and pectoral spines. Due to these constraints, some fish may not even be able to produce a specific sound. In several different species of catfish, aggressive sound production occurs during cover site defence or during threats from other fish.

More specifically, in long-whiskered catfishes, drumming sounds are used as a threatening signal and stridulations are used as a defence signal. Kaatz investigated 83 species from 14 families of catfish, and determined that catfishes produce more stridulatory sounds in disturbance situations and more swimbladder sounds in intraspecific conflicts.

The cat fishes are *predatory fishes*. Their skin is devoid of scales. Two pairs of barbels are present on upper and lower jaw. Most of the fish body is utilized as food due to absence of scales and spines.

Clarias Batracus

It is commonly found in brackish and fresh water ponds of India, South and West Asian countries. Head is slightly compressed and enclosed by plates. Body is brown or dark gray in colour. It is not only used as food but also for experiments in laboratories. It is provided with *Accessory respiratory organ*. It grows to a size of 45 cm.

Heteropneustes Fossilis

The head is flat with laterally compressed body. It possesses accessory respiratory organs and lives in lake kolleru of A.P. It feeds on molluscans, algae and grows to a size of 45 cm.

Clarias Macrocephalus:

It grows to a size of 40cm.

Anabas Testudeneus: (Climbing Perch)

It grows to a length of 15cm and feeds on aquatic insects. Head is triangular with wide mouth and greenish in colour

Etropius Suratensis:

It is commonly called pearl spot due to presence of transparent patches shining like pearls. The body is greenish, light pink and possesses eight black stripes. It is regarded as a good food fish due to delicious smell. It is not a predator, builds nests during the breeding season and hence regarded as most suitable for culture in ponds and lakes.

Wallago Attu

It is a cat fish found in all rivers and lakes of India. Head is larger than trunk.

The body is laterally compressed. Tail is extraordinarily long and slender. It grows to a size of ito 2 mt. but captured at 90 cm. length. Two pairs of barbells are present on the head. Mouth is large with large jaws having teeth for feeding on fresh water. Hence it is called fresh water shark. As it is a predator fish, it is not suitable for artificial culture.

Mystus Seenghala

Four pairs of barbs, elongated upper jaw, long maxillary barbs, deeply divided caudal fin are the main features of the fish.

Patin Fish

Growth Place

This fish is originally from Southeast Asia. It can be found in rivers and lakes of Malaysia, Thailand, Indonesia and others. Patin fish can be found in almost all the major Malaysian water streams such as Pahang River, Kenyir Lake, Kelantan River and others.

Background

Catfish is one of the family members in Patin fish. This fish like to cluster between them and rests in the deeper levels of the water sources. Patin fish has beautiful appearance; its shape is similar to that of the catfish. Small Patin fish are grey coloured and will turn brighter as they turn adult. It is lively fish, often swimming near the water surface. It can breathe outside water as it has strong tolerance to hypoxia. There are two types of body colouration, black and white. Black coloured are also know as Black Patin. In Malaysia, there are four types of Patin fish; River Patin (P.nasutus), Mucung Patin (Helicophagus waandersii), Champion Patin (P. micronemus) and Patin King. Patin King is the result between the combination of Black Patin and River Patin.

Main Food

Omnivorous fish but more specifically belong to carnivorous. It primarily eats aquatic insects, fish, shrimp, all kinds of dead animals, fruit and plant debris.

Fish Sizeÿ

This fish has long and rather flat body, arched back, conical head and small mouth. The gill membrane is not linked to its jaws. It has a large swim bladder that also provides the ability to breathe air. This fish cartilage is similar to that of sharks, whales, squid and eel. It can weigh up to 20 kg.

Aquaculture

Catfish are easy to farm in warm climates, leading to inexpensive and safe food at local grocers. Catfish raised in inland tanks or channels are considered safe for the environment, since their waste and disease should be contained and not spread to the wild.

In Asia, many catfish species are important as food. Several walking catfish (Clariidae) and shark catfish (Pangasiidae) species are heavily cultured in Africa and Asia. Exports of one particular shark catfish species from Vietnam, *Pangasius bocourti*, has met with pressures from the U.S. catfish industry. In 2003, The United States Congress passed a law preventing the imported fish from being labelled as catfish. As a result, the Vietnamese exporters of this fish now label their products sold in the U.S. as "basa fish." Trader Joe's has labelled frozen fillets of Vietnamese *Pangasius hypophthalmus* as "striper."

There is a large and growing ornamental fish trade, with hundreds of species of catfish, such as *Corydoras* and armored suckermouth catfish (often called plecos), being a popular component of many aquaria. Other catfish commonly found in the aquarium trade are banjo catfish, talking catfish, and long-whiskered catfish.

Catfish as Invasive Species

Representatives of the genus *Ictalurus* have been introduced into European waters in the hope of obtaining a sporting and food resource. However, the European stock of American catfishes has not achieved the dimensions of these fish in their native waters, and have only increased the ecological pressure on native European fauna. Walking catfish have also been introduced in the freshwaters of Florida, with the voracious catfish becoming a major alien pest there. Flathead catfish, *Pylodictis olivaris*, is also a North American pest on Atlantic slope drainages. *Pterygoplichthys* species, released by aquarium fishkeepers, have also established feral populations in many warm waters around the world.

Dangers to Humans

While the vast majority of catfish are harmless to humans, a few species are known to present some risk. Many catfish species have "stings" (actually non-venomous in most cases) embedded behind their fins; thus precautions must be taken when handling them.

The Candiru may have a parasitic behaviour, invading the male urethra, although the evidence of this is somewhat inconclusive.

Taxonomy

The catfish are a monophyletic group. This is supported by molecular evidence.

Catfish belong to a superorder called the Ostariophysi, which also includes the Cypriniformes, Characiformes, Gonorynchiformes and Gymnotiformes, a superorder characterized by the Weberian apparatus. Some place Gymnotiformes as a sub-order of Siluriformes, however this is not as widely accepted.

Currently, the Siluriformes are said to be the sister group to the Gymnotiformes, though this has been debated due to more recent molecular evidence. As of 2007 there are about 36 extant catfish families, and about 3,093 extant species have been described. This makes the catfish order the second or third most diverse vertebrate order; in fact, 1 out of every 20 vertebrate species is a catfish.

The taxonomy of catfish is quickly changing. In a 2007 and 2008 paper, *Horabagrus*, *Phreatobius*, and *Conorhynchos* were not classified under any current catfish families. There is disagreement on the family status of certain groups; for example, Nelson (2006) lists Auchenoglanididae and Heteropneustidae as separate families, while the All Catfish Species Inventory (ACSI) includes them under other families. Also, FishBase and the Integrated Taxonomic Information System lists Parakysidae as a separate family, while this group is included under Akysidae by both Nelson (2006) and ACSI. Many sources do not list the recently revised family Anchariidae.

The family Horabagridae, including *Horabagrus*, *Pseudeutropius*, and *Platytropius*, is also not shown by some authors but presented by others as a true group. Thus, the actual number of families differs between authors. The species count is in constant flux due to taxonomic work as well as description of new species. On the other hand, our understanding of catfish should increase in the next few years due to work by the ACSI.

The rate of description of new catfish is at an all-time high. Between 2003 and 2005, over 100 species have been named, a rate three times faster than that of the past century. In June 2005, researchers named the newest family of catfish, Lacantuniidae, only the third new family of fish distinguished in the last 70 years. The new species in

Lacantuniidae, *Lacantunia enigmatica*, was found in the Lacantun river in the Mexican state of Chiapas.

According to morphological data, Diplomystidae is usually considered to be the most primitive of catfish and the sister group to the remaining catfish, grouped in a clade called Siluroidei. Recent molecular evidence contrasts the prevailing hypothesis, where the suborder Loricarioidei are the sister group to all catfish, including Diplomystidae (Diplomystoidei) and Siluroidei; though they were not able to reject the past hypothesis, the new hypothesis is not unsupported. Siluroidei was found to be monophyletic without Loricarioid families or Diplomystidae with molecular evidence; morphological evidence is unknown that supports Siluroidei without Loricarioidea.

Below is a list of family relationships by different authors. Lacantuniidae is included in the Sullivan scheme based on recent evidence that places it sister to Claroteidae.

Chapter 6

Fish Kill

The term fish kill, known also as fish die-off, is a localized die-off of fish populations which may also be associated with more generalized mortality of aquatic life. The most common cause is reduced oxygen in the water, which in turn may be due to factors such as drought, algae bloom, overpopulation, or a sustained increase in water temperature. Infectious diseases and parasites can also lead to fish kill. Toxicity is a real but far less common cause of fish kill.

Fish kills are often the first visible signs of environmental stress and are usually investigated as a matter of urgency by environmental agencies to determine the cause of the kill. Many fish species have a relatively low tolerance of variations in environmental conditions and their death is often a potent indicator of problems in their environment that may be affecting other animals and plants and may have a direct impact on other uses of the water such as for drinking water production. Pollution events may affect fish species and fish age classes in different ways. If it is a cold-related fish kill, juvenile fish or species that are not cold-tolerant may be selectively affected. If toxicity is the cause, species are more generally affected and the event may include amphibians and shellfish as well. A reduction in dissolved oxygen may affect larger specimens more than smaller fish as these may be able to access oxygen richer water at the surface, at least for a short time.

Causes

Fish kills may result from a variety of causes. Of known causes, fish kills are most frequently caused by pollution from agricultural runoff or biotoxins. Ecological hypoxia (oxygen depletion) is one of the most common natural causes of fish kills. The hypoxic event may be brought on by factors such as algae blooms, droughts, high temperatures and thermal pollution. Fish kills may also occur due to the presence of disease, agricultural and sewage runoff, oil or hazardous waste spills, hydraulic fracturing wastewater, sea-quakes, inappropriate re-stocking of fish, poaching with chemicals, underwater explosions, and other catastrophic events that upset a normally stable aquatic population. Because of the difficulty and lack of standard protocol to investigate fish kills, many fish kill cases are designated as having an 'unknown' cause.

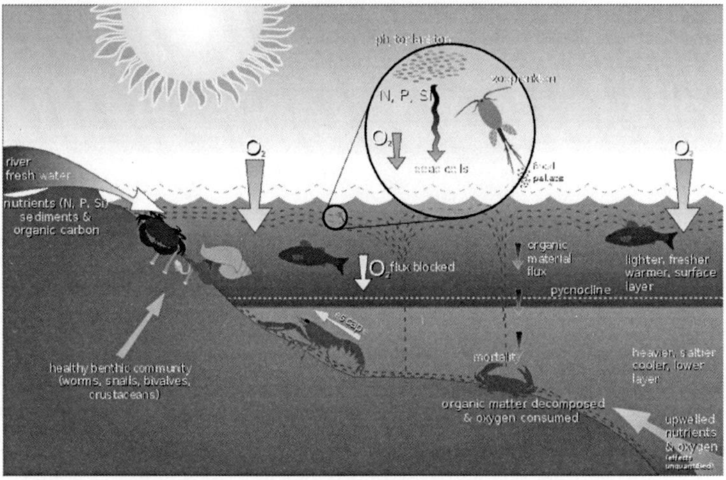

Figure: Depleted oxygen levels are the most common cause of fish kills. In this way eutrophication can have devastating consequences for the health of benthic life

Oxygen Depletion

Oxygen enters the water through diffusion. The amount of oxygen that can be dissolved in water depends on the atmospheric pressure, the water temperature and whether the water is salty. For example, at 20 °C (68 °F) and one atmosphere of pressure, a maximum of 8 mg/l of oxygen can dissolve in sea water (35 mg/l salinity) while a maximum of 9 mg/l of oxygen can dissolve in fresh water. The amount of oxygen that can be dissolved in the water decreases by about 1 mg/l for each 10 °C increase in water temperature above 20 °C.

Many cold water fish that have evolved to live in clean cold waters become stressed when oxygen concentrations fall below 8 mg/l whilst warm water fish generally need at least 5 ppm (5 mg/l) of dissolved oxygen. Fish can endure short periods of reduced oxygen. Depleted oxygen levels are the most common cause of fish kills. Oxygen levels normally fluctuate even over the course of a day and are affected by weather, temperature, the amount of sunlight available, and the amount of living and dead plant and animal matter in the water. In temperate zones oxygen levels in eutrophic rivers in summertime can exhibit very large diurnal fluctuations with many hours of oxygen supersaturation during daylight followed by oxygen depletion at night. Associated with these photosynthetic rhythms there is a matching pH rhythm as bicarbonate ion is metabolised by plant cells. This can lead to pH stress even when oxygen levels are high.

Additional dissolved organic loads are the most common cause of oxygen depletion and such organic loads may come from sewage, farm waste, tip leachate and many other sources

Diseases and Parasites

Fish are subject to various viruses, bacteria and fungi in addition to parasites such as protozoans, flukes and worms, or crustaceans. These are naturally occurring in many bodies of water, and fish that are stressed for other reasons, such as spawning or suboptimal water quality, are more susceptible. Signs of disease include sores, missing scales or lack of slime, strange growths or visible parasites, and abnormal behaviour – lazy, erratic, gasping at the water surface or floating head, tail or belly up.

For example, since 2004 fish kills have been observed in the Shenandoah River basin in the spring, from the time water temperatures are in the 50s (°F) until they reach the mid-70s. So far, investigators suspect certain bacteria, along with environmental and contaminant factors that may cause immune suppression.

In fish farming, where populations are optimized for the available resources, parasites or disease can spread quickly. In channel catfish aquaculture ponds, for example, the "hamburger gill disease" is caused by a protozoan called *Aurantiactinomyxon* and can kill all the fish in an affected pond. In addition to altered behaviour, affected fish have swollen gills that are mottled and have the appearance of ground hamburger meat.

Some early warning signs of fish suffering from disease or parasite infections include:

1. Discolouration, open sores, reddening of the skin, bleeding, black or white spots on the skin
2. Abnormal shape, swollen areas, abnormal lumps, or popeyes
3. Abnormal distribution of the fish such as crowding at the surface, inlet, or pond edges (though crowding at the surface during specific times of day, such as early morning, is more likely a sign of low oxygen)

4. Abnormal activity such as flashing, twisting, whirling, convulsions, loss of buoyancy
5. Listlessness, weakness, sluggishness, lack of activity
6. Loss of appetite or refusal to feed

Toxins

Agricultural runoff, sewage, surface runoff, chemical spills, hazardous waste spills can all potentially lead to water toxicity and fish kill. Some algae species also produce toxins. In Florida, these include *Aphanizomenon*, *Anabaena* and *Microcystis*. Some notable fish kills in Louisiana in the 1950s were due to a specific pesticide called endrin.

Natural instances of toxic conditions can occur, especially in poorly buffered water. Aluminium compound can cause complete fish kills, sometimes associated with autumn turn-over of lakes leading to complex chemical interactions between pH, calcium ions and complex polymeric salts of Aluminium

Human-induced fish kills are unusual, but occasionally a spilled substance causes direct toxicity or a shift in water temperature or pH that can lead to fish kill. For example, in 1997 a phosphate plant in Mulberry, Florida, accidentally dumped 60 million gallons of acidic process water into Skinned Sapling Creek, reducing the pH from about 8 to less than 4 along 36 miles of creek, resulting in the death of about 1.3 million fish.

It is often difficult or impossible to determine whether a potential toxin is the direct cause of a fish kill. For example, hundreds of thousands of fish died after an accidental spill of bourbon whiskey into the Kentucky River near Lawrenceburg. However, officials could not determine whether the fish kill was due to the bourbon directly or to oxygen depletion that resulted when aquatic microbes rapidly began to consume and digest the liquor.

Cyanide is a particular toxic compound that has been used to poach fish. In cyanide poisoning the gills turn a distinctive cherry red. Chlorine introduced as alkaline hypochlorite solution is also extremely toxic leaving pale mucilaginous gills and an over-production of mucilage across the whole body. Lime produces similar symptoms but is also often associated with milk eyes.

Algae Blooms and Red Tides

An algae bloom is the appearance of a large amount of algae or scum floating on the surface of a body of water. Algae blooms are a natural occurrence in nutrient-rich lakes and rivers, though sometimes increased nutrient levels leading to algae blooms are due to fertilizer or animal waste runoff. A few species of algae produce toxins, but most fish kills due to algae bloom are a result of decreased oxygen levels. When the algae die, decomposition uses oxygen in the water that would be available to fish. A fish kill in a lake in Estonia in 2002 was attributed to a combination of algae bloom and high temperatures. When people manage algae blooms in fish ponds, it is recommended that treatments be staggered to avoid too much algae dying at once, which may result in a large drop in oxygen content.

Some diseases result in mass die offs. One of the more bizarre and recently discovered diseases produces huge fish kills in shallow marine waters. It is caused by the ambush predator dinoflagellate *Pfiesteria piscicida*. When large numbers of fish, like shoaling forage fish, are in confined situations such as shallow bays, the excretions from the fish encourage this dinoflagellate, which is not normally toxic, to produce free-swimming zoospores.

If the fish remain in the area, continuing to provide nourishment, then the zoospores start secreting a neurotoxin. This toxin results in the fish developing bleeding lesions, and their skin flakes off in the water. The dinoflagellates then eat the blood and flakes of tissue while the affected fish die. Fish kills by this dinoflagellate are common, and they may also have been responsible for kills in the past which were thought to have had other causes. Kills like these can be viewed as natural mechanisms for regulating the population of exceptionally abundant fish. The rate at which the kills occur increases as organically polluted land runoff increases.

Red tide is the name commonly given to an algal bloom of *Karenia brevis*, a microscopic marine dinoflagellate which is common in Gulf of Mexico waters. In high concentrations it discolours the water which often appears reddish-brown in colour. It produces a toxin which paralyses the central nervous system of fish so they cannot breathe. Dead fish wash up on beaches around Texas and Florida. Humans can also become seriously ill from eating oysters and other shellfish contaminated with the red tide toxin. The term "red tide" is also commonly used to describe harmful algal blooms on the northern east coast of the United States, particularly in the Gulf of Maine. This type of bloom is caused by another species of dinoflagellate known as *Alexandrium fundyense*. These blooms are natural phenomenon, but the exact cause or combination of factors that result in red tides outbreak not fully understood.

Biological Decay

Just as an algae bloom can lead to oxygen depletion, introduction of a large amount of decaying biological material in general to a body of water leads to oxygen depletion as microorganisms use up available oxygen in the process of breaking down organic matter. For example, a 10-mile-long fish kill in September, 2010, in the Sangamon River in Illinois was traced to discharge of animal waste into the river from a large dairy operation. The illegal discharge resulted in a complete kill of fish, frogs, mussels and mudpuppies.

Spawning Fatalities

Figure: *A salmon which has died after spawning*

Some species of fish exhibit mass simultaneous mortality as part of their natural life cycle. Fish kill due to spawning fatalities can occur when fish are exhausted from spawning activities such as courtship, nest building, and the release of eggs or milt (sperm). Fish are generally weaker after spawning and are less resilient than usual to smaller changes in the environment. Examples include the Atlantic salmon and the Sockeye salmon where many of the females routinely die immediately after spawning.

Water Temperature

A fish kill can occur with rapid fluctuations in temperature or sustained high temperatures. Generally, cooler water has the potential to hold more oxygen, so a period of sustained high temperatures can lead to decreased dissolved oxygen in a body of water. An August, 2010, fish kill in Delaware Bay was attributed to low oxygen as a result of high temperatures.

A massive (hundreds of thousands) fish kill at the mouth of the Mississippi River in Louisiana, September, 2010, was attributed to a combination of high temperatures and low tide. Such kills are known to happen in this region in late summer and early fall, but this one was unusually large.

A short period of hot weather can increase temperatures in the surface layer of water, as the warmer water tends to stay near the surface and be further heated by the air. In this case, the top warmer layer may have more oxygen than the lower, cooler layers because it has constant access to atmospheric oxygen. If a heavy wind or cold rain then occurs (usually during the autumn but sometimes in summer), the layers can mix. If the volume of low oxygen water is much greater than the volume in the warm surface layer, this mixing can reduce oxygen levels throughout the water column and lead to fish kill.

Fish kills can also result from a dramatic or prolonged drop in air (and thus, water) temperature. This kind of fish kill is selective – usually the dead fish are species that cannot tolerate cold. This has been observed in cases where a fish native to a more tropical region has been introduced to cooler waters, such as the introduction of the tilapia to bodies of water in Florida. Native to Africa's Nile River, the tilapia stop feeding when water temperatures drop below 60 °F (16 °C) and die when it reaches 45 °F (7 °C). Thus, tilapia that have survived and successfully reproduced in Florida are occasionally killed by a winter cold front.

In January, 2011, a selective fish kill affecting an estimated 2 million juvenile spot fish was attributed to a combination of cold stress and overpopulation after a particularly large spawn.

Underwater Explosions

Underwater explosions can lead to fish kill, and fish with swim bladders are more susceptible. Sometimes underwater explosions are used on purpose to induce fish kills, a generally illegal practice known as blast fishing. Underwater explosions may be accidental or planned, such as for construction, seismic testing, mining or blast testing of structures under water. In many places, an assessment of potential effects of underwater explosions on marine life must be completed and preventive measures taken before blasting.

Droughts and Overstocking

Droughts and overstocking can also result in inland fish kills.

A drought can lead to lower water volumes so that even if the water contains a high level of dissolved oxygen, the reduced volume may not be enough for the fish population. Droughts often occur in conjunction with high temperatures so that the oxygen carrying capacity of the water may also be reduced. Low river flows also reduce the available dilution for permitted discharges of treated sewage or industrial waste. The reduced dilution increases the organic demand for oxygen further reducing the oxygen concentration available to fish

Overstocking of fish (or an unusually large spawn) can also result in inland fish kills. Fish kill due to insufficient oxygen is really a matter of too much demand and too little supply for whatever reason(s). Recommended stocking densities are available from many sources for bodies of water ranging from a home aquarium or backyard pond to commercial aquaculture facilities.

Estimation

Estimating the magnitude of a kill presents a number of problems.

1. Polluted waters are often very turbid or have low transparency making it difficult or impossible to see fish that have sunk
2. Rivers and streams can move fish downstream out of the investigation area.
3. Small fish and fry can decompose or become buried in sediments very quickly and are lost from the count.
4. Predators and scavengers remove and eat fish.
5. Stressed fish may swim up tributaries and die there
6. Many kills are reported only when dead fish resurface due to decompositional gas formation, often several hours after the kill has occurred.

Some very large fish kills may never be estimated because of these factors. The discharge of red aluminium sludge from a reservoir in Hungary into the Marcai River is acknowledged as causing environmental devastation, The loss of adult fish also can have long term impacts on the success of the fishery as the following year's spawning stock may have been lost and recovery of the pre-kill population may take years. The loss of food supplies or recreational income may be very significant to the local economy.

Prevention and Investigation

Fish kills are difficult to predict. Even when conditions that contribute to fish kill are known to exist, prevention is hard because often conditions cannot be improved and fish cannot be safely removed in time. In small ponds, mechanical aeration and/or removal of decaying matter (such as fallen leaves or dead algae) may be reasonable and effective preventive measures.

Many countries in the developed world have specific provisions in place to encourage the public to report fish kills so that a proper investigation can take place. Investigation of the cause of a kill requires a multi-disciplinary approach including on-site environmental measurements, investigation of inputs, review of meteorology and past history, toxicology, fish autopsy, invertebrate analysis and a robust knowledge of the area and its problems.

Notable Events

The counts given below are all estimates. They tend to be underestimates, and may omit, for example, small fish, those removed by scavengers and those that settle to the bottom.

Event/Location	Date	Count	Species	Remarks
Gulf of Mexico(Corpus Christi)	1935	22,000,000		Caused by red tide. This event caused coughing, sneezing and watery red eyes in humans.
River Aeron	1974	10,000	salmon,trout	Discharge ofcreamery waste through poorly maintained sewer. Successful prosecution followed.
River Neath	1976	50,000	salmon,trout	Extreme drought left fish stranded in stagnant pools into which sewers drained.
River Ogmore	1979	50,000	salmon,trout	Spillage of Kymenefrom a paper mill on the River Llynfi a tributary of the Ogmore. Successful prosecution followed and substantial compensation.
Gulf of Mexico	1986	22,000,000	Gulf menhaden,striped mullet, various other species	Caused by red tide.
Rhine River	1986 01	500,000		Caused by spill from Swiss chemical warehouse
Texas coast	1997–1998	21,000,000		Caused by a bloom of *Karenia brevis*
White River; West Fork, Indiana	1999	4,800,000		Caused by an automotive parts maker in Anderson, Indiana, which had discharged 10,000 gallons of the chemical HMP 2000 into the river.
River Dee (United Kingdom)	2000 07	100,000	salmon,trout, perch	Unconfirmed link to release of whey into river
Klamath River	2002 09	70,000	salmon	Low flow of water due to drought and water diversions for agriculture led to heated and shallow water, increasing vulnerability to a gill disease.
Neuse River, North Carolina	2004 09	1,900,000	menhaden	"Natural upwelling" of an acknowledged polluted river. Hydrogen sulfide smell reported
Taal Lake, Luzon, Philippines	2008 01 05	50 metric tons	tilapia	May be linked tovolcanic activity and large fish farms
Liuxihe RiverGuangzhouPeople's Republic of China	2008 09 09	10,000	carp	Unknown
Beaches atThanet, Kent, England	2010 01	20,000	velvet crab	20000 + dead crabs – along with dead starfish, lobsters, sponges and anemones. Probably killed byhypothermia.
Ting River FujianPeople's Republic of China	2010 07	>1,000,000Enough to feed 70,000 people a year		Part of the Zijin mining disaster
Mississippi River;Plaquemines Parish, Louisiana	2010 09	100,000	redfish,trout,flounder	
Arkansas River;Ozark, Arkansas	2010 12	100,000	freshwater drum	Coincided with death of 5,000 Red-winged Blackbirds that fell from the sky.
Chesapeake Bay	2011 01	2,000,000	spot croakers	Included some juvenile croakers. Cold water stress was believed to be the cause.
Jiaxing Xiuzhou District People's Republic of China	2011 01 06	250,000	bream,carp,murrel,silver carp,grass carp	Fish caught and transported to market held in large fish tanks fed with river water. Very rapid die-off and loss exceeded 100 tonnes. Only fish caught from a river under China National Highway 320 east died.
Redondo Beach, California	2011 03	millions	anchovies,mackerel,sardinesan d other small fish	Caused by oxygen deprivation
Taal Lake, Batangas, Philippines	2011 05 29	750 metric tons	Tilapia,milkfish	Caused by oxygen deprivation and largefish farms
Lingayen Gulf, Anda, Pangasinan, Philippines	2011 05 30	500 metric tons	Milkfish	Oxygen depletion and change of water climate
Nordreisa, Troms, Norway	2011 12 31	several tons	herring	
Guangxi, People's Republic of China	2012 01 15	40,000 kilograms	Various	Caused by 2012 Guangxi cadmium spill

Aquatic Biomonitoring

Aquatic biomonitoring is the science of inferring the ecological condition of rivers, lakes, streams, and wetlands by examining the organisms that live there. While aquatic biomonitoring is the most common form of such biomonitoring, any ecosystem can be studied in this manner.

Biomonitoring typically takes different approaches:

- *Bioassays*, where test organisms are exposed to an environment to see if mutations or deaths occur. Typical organisms used in bioassays are fish, water fleas (Daphnia), and frogs.
- *Community assessments*, also called *biosurveys,* where an entire community of organisms is sampled, to see what types of taxa remain. In aquatic ecosystems, these assessments often focus on invertebrates, algae, macrophytes (aquatic plants), fish, or amphibians. Rarely, other large vertebrates (reptiles, birds, and mammals) are considered as well.
- *Online biomonitoring devices*, using the ability of animals to permanently taste their environment. Different types of animals are used for that purpose either under lab or field conditions. The use of valve opening/closing activity of clams is one of the possible ways to monitor in-situ the quality of freshwater and coastal waters.

Aquatic invertebrates have the longest history of use in biomonitoring programmes. In typical unpolluted temperate streams of Europe and North America, certain insect taxa predominate. Mayflies (Ephemeroptera), caddisflies (Trichoptera), and stoneflies (Plecoptera) are the most common insects in these undisturbed streams. In rivers disturbed by urbanization, agriculture, forestry, and other perturbations, flies (Diptera), and especially midges (family Chironomidae) predominate. Aquatic invertebrates are responsive to climate change.

Fish Physiology

Fish physiology is the scientific study of how the component parts of fish function together in the living fish. It can be contrasted with fish anatomy, which is the study of the form or morphology of fishes. In practice, fish anatomy and physiology complement each other, the former dealing with the structure of a fish, its organs or component parts and how they are put together, such as might be observed on the dissecting table or under the microscope, and the later dealing with how those components function together in the living fish.

Figure: *When threatened, the toxic pufferfish fills its extremely elastic stomach with water.*

Respiration

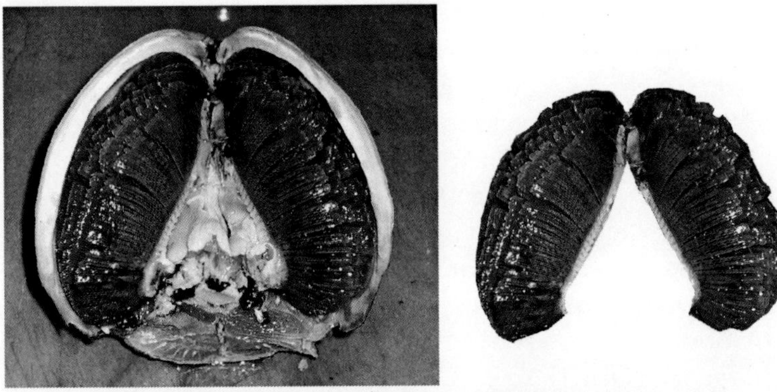

Figure: *Gills inside the head of a tuna. The head is snout-down, with the view looking towards the mouth. On the right are the detached gills.*

Figure: *Gill arches bearing gills in a pike*

Most fish exchange gases using gills on either side of the pharynx (throat). Gills are tissues which consist of threadlike structures called filaments. These filaments have many functions and "are involved in ion and water transfer as well as oxygen, carbon dioxide, acid and ammonia exchange. Each filament contains a capillary network that provides a large surface area for exchanging oxygen and carbon dioxide. Fish exchange gases by pulling oxygen-rich water through their mouths and pumping it over their gills. In some fish, capillary blood flows in the opposite direction to the water, causing countercurrent exchange. The gills push the oxygen-poor water out through openings in the sides of the pharynx.

Fish from multiple groups can live out of the water for extended time periods. Amphibious fish such as the mudskipper can live and move about on land for up to several days, or live in stagnant or otherwise oxygen depleted water. Many such fish can breathe air via a variety of mechanisms. The skin of anguillid eels may absorb oxygen directly. The buccal cavity of the electric eel may breathe air. Catfish of the families Loricariidae, Callichthyidae, and Scoloplacidae absorb air through their digestive tracts. Lungfish, with the exception of the Australian lungfish, and bichirs have paired lungs similar to those of tetrapods and must surface to gulp fresh air through the mouth and pass spent air out through the gills. Gar and bowfin have a vascularized swim bladder that functions in the same way. Loaches, trahiras, and many catfish breathe by passing air through the gut. Mudskippers breathe by absorbing oxygen across the skin (similar to frogs). A number of fish have evolved so-called accessory breathing organs that extract oxygen from the air. Labyrinth fish (such as gouramis and bettas) have a labyrinth organ above the gills that performs this function. A few other fish have structures resembling labyrinth organs in form and function, most notably snakeheads, pikeheads, and the Clariidae catfish family.

Breathing air is primarily of use to fish that inhabit shallow, seasonally variable waters where the water's oxygen concentration may seasonally decline. Fish dependent solely on dissolved oxygen, such as perch and cichlids, quickly suffocate, while air-breathers survive for much longer, in some cases in water that is little more than wet mud. At the most extreme, some air-breathing fish are able to survive in damp burrows for weeks without water, entering a state of aestivation (summertime hibernation) until water returns.

Air breathing fish can be divided into *obligate* air breathers and *facultative* air breathers. Obligate air breathers, such as the African lungfish, are obligated to breathe air periodically or they suffocate. Facultative air breathers, such as the catfish *Hypostomus plecostomus*, only breathe air if they need to and can otherwise rely on their gills for oxygen. Most air breathing fish are facultative air breathers that avoid the energetic cost of rising to the surface and the fitness cost of exposure to surface predators.

All basal vertebrates breathe with gills. The gills are carried right behind the head, bordering the posterior margins of a series of openings from the esophagus to the exterior. Each gill is supported by a cartilagenous or bony gill arch. The gills of vertebrates typically develop in the walls of the pharynx, along a series of gill slits opening to the exterior. Most species employ a countercurrent exchange system to enhance the diffusion of substances in and out of the gill, with blood and water flowing in opposite directions to each other.

The gills are composed of comb-like filaments, the gill lamellae, which help increase their surface area for oxygen exchange. When a fish breathes, it draws in a mouthful of water at regular intervals. Then it draws the sides of its throat together, forcing the water through the gill openings, so that it passes over the gills to the outside. The bony fish have three pairs of arches, cartilaginous fish have five to seven pairs, while the primitive jawless fish have seven. The vertebrate ancestor no doubt had more arches, as some of their chordate relatives have more than 50 pairs of gills.

Higher vertebrates do not develop gills, the gill arches form during fetal development, and lay the basis of essential structures such as jaws, the thyroid gland, the larynx, the *columella* (corresponding to the stapes in mammals) and in mammals the malleus and incus. Fish gill slits may be the evolutionary ancestors of the tonsils, thymus gland, and Eustachian tubes, as well as many other structures derived from the embryonic branchial pouches.

Scientists have investigated what part of the body is responsible for maintaining the respiratory rhythm. They found that neurons located in the brainstem of fish are responsible for the genesis of the respiratory rhythm. The position of these neurons is slightly different from the centres of respiratory genesis in mammals but they are located in the same brain compartment, which has caused debates about the homology of respiratory centres between aquatic and terrestrial species. In both aquatic and terrestrial respiration, the exact mechanisms by which neurons can generate this involuntary rhythm are still not completely understood.

Another important feature of the respiratory rhythm is that it is modulated to adapt to the oxygen consumption of the body. As observed in mammals, fish "breathe" faster and heavier when they do physical exercise. The mechanisms by which these changes occur have been strongly debated over more than 100 years between scientists. The authors can be classified in 2 schools:

1. Those who think that the major part of the respiratory changes are pre-programmed in the brain, which would imply that neurons from locomotion centres of the brain connect to respiratory centres in anticipation of movements.
2. Those who think that the major part of the respiratory changes result from the detection of muscle contraction, and that respiration is adapted as a consequence of muscular contraction and oxygen consumption. This would imply that the brain possesses some kind of detection mechanisms that would trigger a respiratory response when muscular contraction occurs.

Many now agree that both mechanisms are probably present and complementary, or working alongside a mechanism that can detect changes in oxygen and/or carbon dioxide blood saturation.

Bony Fish

In bony fish, the gills lie in a branchial chamber covered by a bony operculum. The great majority of bony fish species have five pairs of gills, although a few have lost some over the course of evolution. The operculum can be important in adjusting the pressure of water inside of the pharynx to allow proper ventilation of the gills, so that bony fish do not have to rely on ram ventilation (and hence near constant motion) to breathe. Valves inside the mouth keep the water from escaping.

The gill arches of bony fish typically have no septum, so that the gills alone project from the arch, supported by individual gill rays. Some species retain gill rakers. Though all but the most primitive bony fish lack a spiracle, the pseudobranch associated with it often remains, being located at the base of the operculum. This is, however, often greatly reduced, consisting of a small mass of cells without any remaining gill-like structure.

Marine teleosts also use gills to excrete electrolytes. The gills' large surface area tends to create a problem for fish that seek to regulate the osmolarity of their internal fluids. Saltwater is less dilute than these internal fluids, so saltwater fish lose large quantities of water osmotically through their gills. To regain the water, they drink large amounts of seawater and excrete the salt. Freshwater is more dilute than the internal fluids of fish, however, so freshwater fish gain water osmotically through their gills.

In some primitive bony fishes and amphibians, the larvae bear external gills, branching off from the gill arches. These are reduced in adulthood, their function taken over by the gills proper in fishes and by lungs in most amphibians. Some amphibans retain the external larval gills in adulthood, the complex internal gill system as seen in fish apparently being irrevocably lost very early in the evolution of tetrapods.

Cartilaginous Fish

Like other fish, sharks extract oxygen from seawater as it passes over their gills. Unlike other fish, shark gill slits are not covered, but lie in a row behind the head. A modified slit called a spiracle lies just behind the eye, which assists the shark with taking in water during respiration and plays a major role in bottom–dwelling sharks. Spiracles are reduced or missing in active pelagic sharks. While the shark is moving, water passes through the mouth and over the gills in a process known as "ram ventilation". While at rest, most sharks pump water over their gills to ensure a constant supply of oxygenated water. A small number of species have lost the ability to pump water through their gills and must swim without rest. These species are *obligate ram ventilators* and would presumably asphyxiate if unable to move. Obligate ram ventilation is also true of some pelagic bony fish species.

The respiration and circulation process begins when deoxygenated blood travels to the shark's two-chambered heart. Here the shark pumps blood to its gills via the ventral aorta artery where it branches into afferent brachial arteries. Reoxygenation takes place in the gills and the reoxygenated blood flows into the efferent brachial arteries, which come together to form the dorsal aorta. The blood flows from the dorsal aorta throughout the body. The deoxygenated blood from the body then flows through the posterior cardinal veins and enters the posterior cardinal sinuses. From there blood enters the heart ventricle and the cycle repeats.

Sharks and rays typically have five pairs of gill slits that open directly to the outside of the body, though some more primitive sharks have six or seven pairs. Adjacent slits are separated by a cartilaginous gill arch from which projects a long sheet-like septum, partly supported by a further piece of cartilage called the *gill ray*. The individual lamellae of the gills lie on either side of the septum. The base of the arch may also support gill rakers, small projecting elements that help to filter food from the water.

A smaller opening, the spiracle, lies in the back of the first gill slit. This bears a small *pseudobranch* that resembles a gill in structure, but only receives blood already oxygenated by the true gills. The spiracle is thought to be homologous to the ear opening in higher vertebrates.

Most sharks rely on ram ventilation, forcing water into the mouth and over the gills by rapidly swimming forward. In slow-moving or bottom dwelling species, especially among skates and rays, the spiracle may be enlarged, and the fish breathes by sucking water through this opening, instead of through the mouth.

Chimaeras differ from other cartilagenous fish, having lost both the spiracle and the fifth gill slit. The remaining slits are covered by an operculum, developed from the septum of the gill arch in front of the first gill.

Lampreys and Hagfish

Lampreys and hagfish do not have gill slits as such. Instead, the gills are contained in spherical pouches, with a circular opening to the outside. Like the gill slits of higher fish, each pouch contains two gills. In some cases, the openings may be fused together, effectively forming an operculum. Lampreys have seven pairs of pouches, while hagfishes may have six to fourteen, depending on the species. In the hagfish, the pouches connect with the pharynx internally. In adult lampreys, a separate respiratory tube develops beneath the pharynx proper, separating food and water from respiration by closing a valve at its anterior end.

Circulation

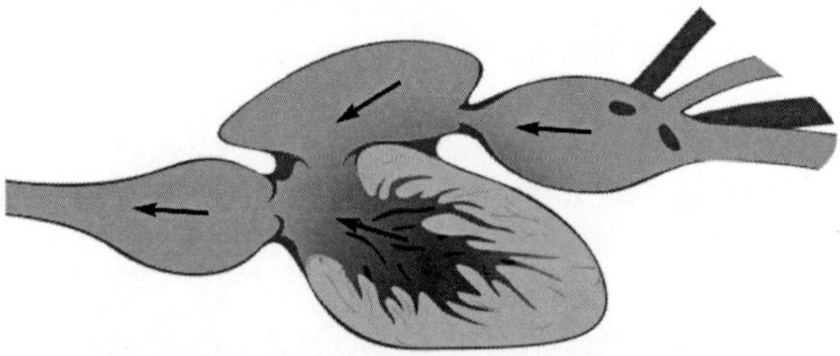

Figure: *Two-chambered heart of a fish*

The circulatory systems of all vertebrates are *closed*, just as in humans. Still, the systems of fish, amphibians, reptiles, and birds show various stages of the evolution of the circulatory system. In fish, the system has only one circuit, with the blood being pumped through the capillaries of the gills and on to the capillaries of the body tissues. This is known as *single cycle* circulation. The heart of fish is therefore only a single pump (consisting of two chambers). Fish have a closed-loop circulatory system. The heart pumps the blood in a single loop throughout the body. In most fish, the heart consists of four parts, including two chambers and an entrance and exit. The first part is the sinus venosus, a thin-walled sac that collects blood from the fish's veins before allowing it to flow to the second part, the atrium, which is a large muscular chamber. The atrium serves as a one-way antechamber, sends blood to the third part, ventricle. The ventricle is another thick-walled, muscular chamber and it pumps the blood, first to the fourth part, bulbus arteriosus, a large tube, and then out of the heart. The bulbus arteriosus connects to the aorta, through which blood flows to the gills for oxygenation.

In amphibians and most reptiles, a double circulatory system is used, but the heart is not always completely separated into two pumps. Amphibians have a three-chambered heart.

Digestion

Jaws allow fish to eat a wide variety of food, including plants and other organisms. Fish ingest food through the mouth and break it down in the esophagus. In the stomach, food is further digested and, in many fish, processed in finger-shaped pouches called pyloric caeca, which secrete digestive enzymes and absorb nutrients. Organs such as the liver and pancreas add enzymes and various chemicals as the food moves through the digestive tract. The intestine completes the process of digestion and nutrient absorption.

In most vertebrates, digestion is a four-stage process involving the main structures of the digestive tract, starting with ingestion, placing food into the mouth, and concluding with the excretion of undigested material through the anus. From the mouth, the food moves to the stomach, where as bolus it is broken down chemically. It then moves to the intestine, where the process of breaking the food down into simple molecules continues and the results are absorbed as nutrients into the circulatory and lymphatic system.

Although the precise shape and size of the stomach varies widely among different vertebrates, the relative positions of the oesophageal and duodenal openings remain relatively constant. As a result, the organ always curves somewhat to the left before curving back to meet the pyloric sphincter. However, lampreys, hagfishes, chimaeras, lungfishes, and some teleost fish have no stomach at all, with the oesophagus opening directly into the intestine. These animals all consume diets that either require little storage of food, or no pre-digestion with gastric juices, or both.

The small intestine is the part of the digestive tract following the stomach and followed by the large intestine, and is where much of the digestion and absorption of food takes place. In fish, the divisions of the small intestine are not clear, and the terms *anterior* or *proximal* intestine may be used instead of duodenum.

The small intestine is found in all teleosts, although its form and length vary enormously between species. In teleosts, it is relatively short, typically around one and a half times the length of the fish's body. It commonly has a number of *pyloric caeca*, small pouch-like structures along its length that help to increase the overall surface area of the organ for digesting food. There is no ileocaecal valve in teleosts, with the boundary between the small intestine and the rectum being marked only by the end of the digestive epithelium.

There is no small intestine as such in non-teleost fish, such as sharks, sturgeons, and lungfish. Instead, the digestive part of the gut forms a spiral intestine, connecting the stomach to the rectum. In this type of gut, the intestine itself is relatively straight, but has a long fold running along the inner surface in a spiral fashion, sometimes for dozens of turns. This valve greatly increases both the surface area and the effective length of the intestine. The lining of the spiral intestine is similar to that of the small intestine in teleosts and non-mammalian tetrapods. In lampreys, the spiral valve is extremely small, possibly because their diet requires little digestion. Hagfish have no spiral valve at all, with digestion occurring for almost the entire length of the intestine, which is not subdivided into different regions.

The large intestine is the last part of the digestive system normally found in vertebrate animals. Its function is to absorb water from the remaining indigestible food matter, and then to pass useless waste material from the body. In fish, there is no true large intestine, but simply a short rectum connecting the end of the digestive part of the gut to the cloaca. In sharks, this includes a *rectal gland* that secretes salt to help the animal maintain osmotic balance with the seawater. The gland somewhat resembles a caecum in structure, but is not a homologous structure.

As with many aquatic animals, most fish release their nitrogenous wastes as ammonia. Some of the wastes diffuse through the gills. Blood wastes are filtered by the kidneys.

Saltwater fish tend to lose water because of osmosis. Their kidneys return water to the body. The reverse happens in freshwater fish: they tend to gain water osmotically. Their kidneys produce dilute urine for excretion. Some fish have specially adapted kidneys that vary in function, allowing them to move from freshwater to saltwater.

In sharks, digestion can take a long time. The food moves from the mouth to a J-shaped stomach, where it is stored and initial digestion occurs. Unwanted items may never get past the stomach, and instead the shark either vomits or turns its stomachs inside out and ejects unwanted items from its mouth. One of the biggest differences between the digestive systems of sharks and mammals is that sharks have much shorter intestines. This short length is achieved by the spiral valve with multiple turns within a single short section instead of a long tube-like intestine. The valve provides a long surface area, requiring food to circulate inside the short gut until fully digested, when remaining waste products pass into the cloaca.

Osmoregulation

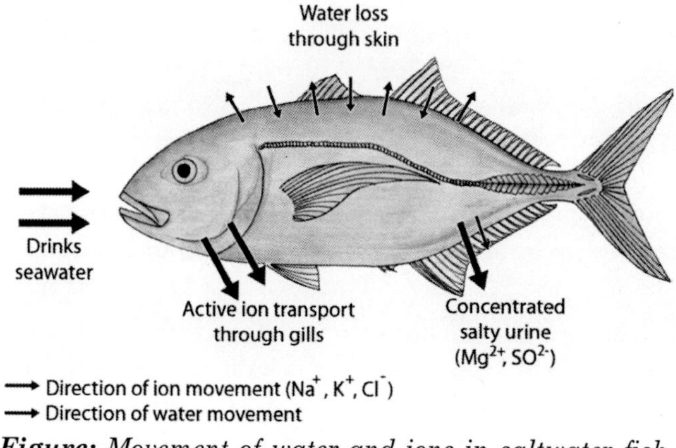

Figure: *Movement of water and ions in saltwater fish*

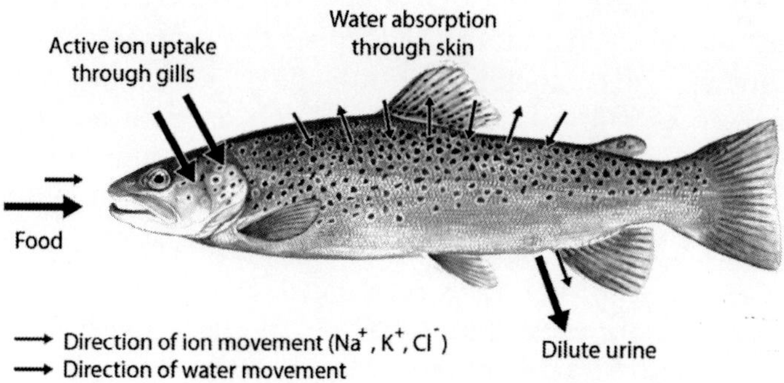

Figure: *Movement of water and ions in freshwater fish*

Two major types of osmoregulation are osmoconformers and osmoregulators. Osmoconformers match their body osmolarity to their environment actively or passively. Most marine invertebrates are osmoconformers, although their ionic composition may be different from that of seawater.

Osmoregulators tightly regulate their body osmolarity, which always stays constant, and are more common in the animal kingdom. Osmoregulators actively control salt concentrations despite the salt concentrations in the environment. An example is freshwater fish. The gills actively uptake salt from the environment by the use of mitochondria-rich cells.

Water will diffuse into the fish, so it excretes a very hypotonic (dilute) urine to expel all the excess water. A marine fish has an internal osmotic concentration lower than that of the surrounding seawater, so it tends to lose water and gain salt. It actively excretes salt out from the gills.

Most fish are stenohaline, which means they are restricted to either salt or fresh water and cannot survive in water with a different salt concentration than they are adapted to. However, some fish show a tremendous ability to effectively osmoregulate across a broad range of salinities; fish with this ability are known as euryhaline species, e.g., salmon. Salmon has been observed to inhabit two utterly disparate environments — marine and fresh water — and it is inherent to adapt to both by bringing in behavioural and physiological modifications.

In contrast to bony fish, with the exception of the coelacanth, the blood and other tissue of sharks and Chondrichthyes is generally isotonic to their marine environments because of the high concentration of urea and trimethylamine N-oxide (TMAO), allowing them to be in osmotic balance with the seawater.

This adaptation prevents most sharks from surviving in freshwater, and they are therefore confined to marine environments. A few exceptions exist, such as the bull shark, which has developed a way to change its kidney function to excrete large amounts of urea. When a shark dies, the urea is broken down to ammonia by bacteria, causing the dead body to gradually smell strongly of ammonia.

Sharks have adopted a different, efficient mechanism to conserve water, i.e., osmoregulation. They retain urea in their blood in relatively higher concentration. Urea is damaging to living tissue so, to cope with this problem, some fish retain *trimethylamine oxide*.

This provides a better solution to urea's toxicity. Sharks, having slightly higher solute concentration (i.e., above 1000 mOsm which is sea solute concentration), do not drink water like fresh water fish.

Thermoregulation

Homeothermy and poikilothermy refer to how stable an organism's temperature is. Most endothermic organisms are homeothermic, like mammals. However, animals with facultative endothermy are often poikilothermic, meaning their temperature can vary considerably. Similarly, most fish are ectotherms, as all of their heat comes from the surrounding water. However, most are homeotherms because their temperature is very stable.

H.M. Vernon has done work on the death temperature and paralysis temperature (temperature of heat rigor) of various animals. He found that species of the same class showed very similar temperature values, those from the Amphibia examined being 38.5 °C, fish 39 °C, Reptilia 45 °C, and various Molluscs 46 °C.

To cope with low temperatures, some fish have developed the ability to remain functional even when the water temperature is below freezing; some use natural antifreeze or antifreeze proteins to resist ice crystal formation in their tissues.

Most sharks are "cold-blooded" or, more precisely, poikilothermic, meaning that their internal body temperature matches that of their ambient environment. Members of the family Lamnidae (such as the shortfin mako shark and the great white shark) are homeothermic and maintain a higher body temperature than the surrounding water. In these sharks, a strip of aerobic red muscle located near the centre of the body generates the heat, which the body retains via a countercurrent exchange mechanism by a system of blood vessels called the rete mirabile ("miraculous net"). The common thresher shark has a similar mechanism for maintaining an elevated body temperature, which is thought to have evolved independently.

Tuna can maintain the temperature of certain parts of their body above the temperature of ambient seawater. For example, bluefin tuna maintain a core body temperature of 25–33 °C (77–91 °F), in water as cold as 6 °C (43 °F). However, unlike typical endothermic creatures such as mammals and birds, tuna do not maintain temperature within a relatively narrow range. Tuna achieve endothermy by conserving the heat generated through normal metabolism. The rete mirabile ("wonderful net"), the intertwining of veins and arteries in the body's periphery, transfers heat from venous blood to arterial blood via a counter-current exchange system, thus mitigating the effects of surface cooling. This allows the tuna to elevate the temperatures of the highly aerobic tissues of the skeletal muscles, eyes and brain, which supports faster swimming speeds and reduced energy expenditure, and which enables them to survive in cooler waters over a wider range of ocean environments than those of other fish. In all tunas, however, the heart operates at ambient temperature, as it receives cooled blood, and coronary circulation is directly from the gills.

- Homeothermy: Although most fish are exclusively ectothermic, there are exceptions. Certain species of fish maintain elevated body temperatures. Endothermic teleosts (bony fish) are all in the suborder Scombroidei and include the billfishes, tunas, including a "primitive" mackerel species, *Gasterochisma melampus*. All sharks in the family Lamnidae – shortfin mako, long fin mako, white, porbeagle, and salmon shark – are endothermic, and evidence suggests the trait exists in family Alopiidae (thresher sharks). The degree of endothermy varies from the billfish, which warm only their eyes and brain, to bluefin tuna and porbeagle sharks who maintain body temperatures elevated in excess of 20 °C above ambient water temperatures. Endothermy, though metabolically costly, is thought to provide advantages such as increased muscle strength, higher rates of central nervous system processing, and higher rates of digestion.

In some fish, a rete mirabile allows for an increase in muscle temperature in regions where this network of vein and arteries is found. The fish is able to thermoregulate certain areas of their body. Additionally, this increase in temperature leads to an increase in basal metabolic temperature. The fish is now able to split ATP at a higher rate and ultimately can swim faster.

The eye of a swordfish can generate heat to better cope with detecting their prey at depths of 2000 feet.

Muscular System

Fish swim by contracting longitudinal red muscle and obliquely oriented white muscles. The red muscle is aerobic and needs oxygen which is supplied by myoglobin. The white muscle is anaerobic and it does not need oxygen. Red muscles are used for sustained activity such as cruising at slow speeds on ocean migrations. White muscles are used for bursts of activity, such as jumping or sudden bursts of speed for catching prey.

Mostly fish have white muscles, but the muscles of some fishes, such as scombroids and salmonids, range from pink to dark red. The red myotomal muscles derive their colour from myoglobin, an oxygen-binding molecule, which tuna express in quantities far higher than most other fish. The oxygen-rich blood further enables energy delivery to their muscles.

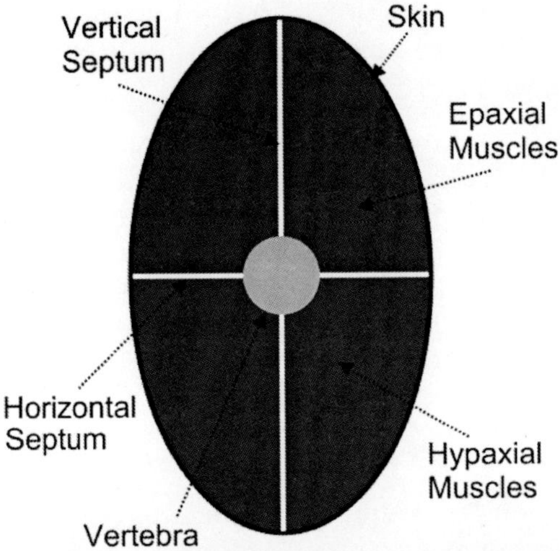

Figure: *Cross section of a teleost*

Figure: *Iridescent shark filets showing myomere structure*

Most fish move by alternately contracting paired sets of muscles on either side of the backbone. These contractions form S-shaped curves that move down the body. As each curve reaches the back fin, backward force is applied to the water, and in conjunction with the fins, moves the fish forward. The fish's fins function like an airplane's flaps. Fins also increase the tail's surface area, increasing speed. The streamlined body of the fish decreases the amount of friction from the water.

A typical characteristic of many animals that utilize undulatory locomotion is that they have segmented muscles, or blocks of myomeres, running from their head to tails which are separated by connective tissue called myosepta. In addition, some segmented muscle groups, such as the of the lateral hypaxial musculature in the salamander are oriented at an angle to the longitudinal direction. For these obliquely oriented fibres the strain in the longitudinal direction is greater than the strain in the muscle fibre direction leading to an architectural gear ratio greater than 1. A higher initial angle of orientation and more dorsoventral bulging produces a faster muscle contraction but results in a lower amount of force production.

It is hypothesized that animals employ a variable gearing mechanism that allows self-regulation of force and velocity to meet the mechanical demands of the contraction. When a pennate muscle is subjected to a low force, resistance to width changes in the muscle cause it to rotate which consequently produce a higher architectural gear ratio (AGR) (high velocity). However, when subject to a high force, the perpendicular fibre force component overcomes the resistance to width changes and the muscle compresses producing a lower AGR (capable of maintaining a higher force output).

Most fishes bend as a simple, homogenous beam during swimming via contractions of longitudinal red muscle fibres and obliquely oriented white muscle fibres within the segmented axial musculature. The fibre strain (εf) experienced by the longitudinal red muscle fibres is equivalent to the longitudinal strain (εx). The deeper white muscle fibres fishes show diversity in arrangement. These fibres are organized into cone-shaped structures and attach to connective tissue sheets known as myosepta; each fibre shows a characteristic dorsoventral (α) and mediolateral (φ) trajectory. The segmented architecture theory predicts that, $\varepsilon x > \varepsilon f$. This phenomenon results in an architectural gear ratio, determined as longitudinal strain divided by fibre strain $(\varepsilon x / \varepsilon f)$, greater than one and longitudinal velocity amplification; furthermore, this emergent velocity amplification may be augmented by variable architectural gearing via mesolateral and dorsoventral shape changes, a pattern seen in pennate muscle contractions. A red-to-white gearing ratio (red εf / white εf) captures the combined effect of the longitudinal red muscle fibre and oblique white muscle fibre strains.

Undulatory Locomotion

Undulatory locomotion is the type of motion characterized by wave-like movement patterns that act to propel an animal forward. Examples of this type of gait include crawling in snakes, or swimming in the lamprey. Although this is typically the type of gait utilized by limbless animals, some creatures with limbs, such as the salamander, choose to forgo use of their legs in certain environments and exhibit undulatory locomotion. This movement strategy is important to study in order to create novel robotic devices capable of traversing a variety of environments.

Buoyancy

The body of a fish is denser than water, so fish must compensate for the difference or they will sink. Many bony fishes have an internal organ called a swim bladder, or gas bladder, that adjusts their buoyancy through manipulation of gases. In this way, fish can stay at the current water depth, or ascend or descend without having to waste energy in swimming. The bladder is only found in bony fishes. In the more primitive groups like some minnows, bichirs and lungfish, the bladder is open to the esophagus and double as a lung. It is often absent in fast swimming fishes such as the tuna and mackerel families. The condition of a bladder open to the esophagus is called physostome, the closed condition physoclist. In the latter, the gas content of the bladder is controlled through the rete mirabilis, a network of blood vessels effecting gas exchange between the bladder an the blood.

In some fish, a rete mirabile fills the swim bladder with oxygen. A countercurrent exchange system is utilized between the venous and arterial capillaries. By lowering the pH levels in the venous capillaries, oxygen unbinds from blood hemoglobin. This causes an increase in venous blood oxygen concentration, allowing the oxygen to diffuse through the capillary membrane and into the arterial capillaries, where oxygen is still sequestered to hemoglobin. The cycle of diffusion continues until the concentration of oxygen in the arterial capillaries is supersaturated (larger than the concentration of oxygen in the swim bladder). At this point, the free oxygen in the arterial capillaries diffuses into the swim bladder via the gas gland.

Unlike bony fish, sharks do not have gas-filled swim bladders for buoyancy. Instead, sharks rely on a large liver filled with oil that contains squalene, and their cartilage, which is about half the normal density of bone. Their liver constitutes up to 30% of their total body mass. The liver's effectiveness is limited, so sharks employ dynamic lift to maintain depth when not swimming. Sand tiger sharks store air in their stomachs, using it as a form of swim bladder. Most sharks need to constantly swim in order to breathe and cannot sleep very long without sinking (if at all). However, certain species, like the nurse shark, are capable of pumping water across their gills, allowing them to rest on the ocean bottom.

Sensory Systems

Most fish possess highly developed sense organs. Nearly all daylight fish have colour vision that is at least as good as a human's. Many fish also have chemoreceptors that are responsible for extraordinary senses of taste and smell. Although they have ears, many fish may not hear very well. Most fish have sensitive receptors that form the lateral line system, which detects gentle currents and vibrations, and senses the motion of nearby fish and prey. Sharks can sense frequencies in the range of 25 to 50 Hz through their lateral line.

Fish orient themselves using landmarks and may use mental maps based on multiple landmarks or symbols. Fish behaviour in mazes reveals that they possess spatial memory and visual discrimination.

Vision

Vision is an important sensory system for most species of fish. Fish eyes are similar to those of terrestrial vertebrates like birds and mammals, but have a more spherical lens. Their retinas generally have both rod cells and cone cells (for scotopic and photopic vision), and most species have colour vision. Some fish can see ultraviolet and some can see polarized light. Amongst jawless fish, the lamprey has well-developed eyes, while the hagfish has only primitive eyespots. Fish vision shows adaptation to their visual environment, for example deep sea fishes have eyes suited to the dark environment.

Hearing

Hearing is an important sensory system for most species of fish. Hearing threshold and the ability to localize sound sources are reduced underwater, in which the speed of sound is faster than in air. Underwater hearing is by bone conduction, and localization of sound appears to depend on differences in amplitude detected by bone conduction. Aquatic animals such as fish, however, have a more specialized hearing apparatus that is effective underwater.

Fish can sense sound through their lateral lines and their otoliths (ears). Some fishes, such as some species of carp and herring, hear through their swim bladders, which function rather like a hearing aid.

Hearing is well-developed in carp, which have the Weberian organ, three specialized vertebral processes that transfer vibrations in the swim bladder to the inner ear.

Although it is hard to test sharks' hearing, they may have a sharp sense of hearing and can possibly hear prey many miles away. A small opening on each side of their heads (not the spiracle) leads directly into the inner ear through a thin channel. The lateral line shows a similar arrangement, and is open to the environment via a series of openings called lateral line pores. This is a reminder of the common origin of these two vibration- and sound-detecting organs that are grouped together as the acoustico-lateralis system. In bony fish and tetrapods the external opening into the inner ear has been lost.

Chemoreception

Sharks have keen olfactory senses, located in the short duct (which is not fused, unlike bony fish) between the anterior and posterior nasal openings, with some species able to detect as little as one part per million of blood in seawater.

Sharks have the ability to determine the direction of a given scent based on the timing of scent detection in each nostril. This is similar to the method mammals use to determine direction of sound.

They are more attracted to the chemicals found in the intestines of many species, and as a result often linger near or in sewage outfalls. Some species, such as nurse sharks, have external barbels that greatly increase their ability to sense prey.

Electroreception

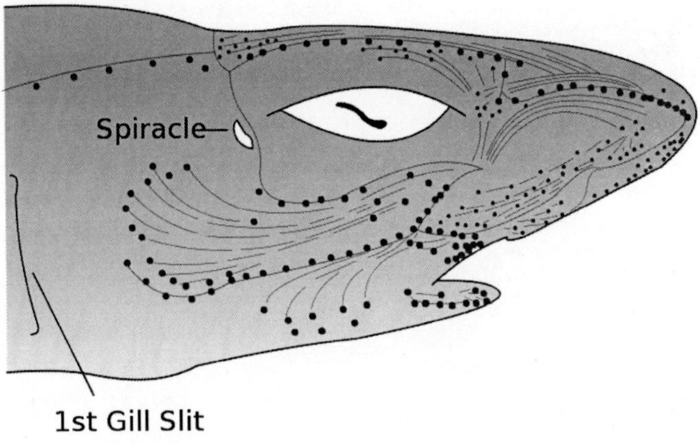

Figure: *Electromagnetic field receptors (ampullae of Lorenzini) and motion detecting canals in the head of a shark*

Some fish, such as catfish and sharks, have organs that detect weak electric currents on the order of millivolt. Other fish, like the South American electric fishes Gymnotiformes, can produce weak electric currents, which they use in navigation and social communication. In sharks, the ampullae of Lorenzini are electroreceptor organs. They number in the hundreds to thousands. Sharks use the ampullae of

Lorenzini to detect the electromagnetic fields that all living things produce. This helps sharks (particularly the hammerhead shark) find prey. The shark has the greatest electrical sensitivity of any animal. Sharks find prey hidden in sand by detecting the electric fields they produce. Ocean currents moving in the magnetic field of the Earth also generate electric fields that sharks can use for orientation and possibly navigation.

- The ampullae of Lorenzini allow sharks to sense electrical discharges.
- Electric fish are able to produce electric fields by modified muscles in their body.

Electrocommunication

Electrocommunication is the communication method used by weakly electric fishes. Weakly electric fishes are a group of animals that utilize a communicating channel that is "invisible" to most other animals: electric signalling. Electric fishes communicate electrically by one fish generating an electric field and a second individual receiving that electric field with its electroreceptors.

The receiving side will interpret the signal frequencies, waveforms, and delay, etc. The best studied species are two freshwater lineages-the African Mormyridae and the South American Gymnotiformes. While weakly electric fish are the only group that have been identified to carry out both generation and reception of electric fields, other species either generate signals or receive them, but not both.

Animals that either generate or receive electric fields are found only in aquatic (or at least moist) environments due to large resistance of all other media (e.g. air). So far, communication between electric fish has been identified mainly to serve the purpose of conveying information in

- species recognition
- courtship and sex recognition
- motivational status (attack warning or submission) and
- environmental conditions.

Pain

Experiments done by William Tavolga provide evidence that fish have pain and fear responses. For instance, in Tavolga's experiments, toadfish grunted when electrically shocked and over time they came to grunt at the mere sight of an electrode.

In 2003, Scottish scientists at the University of Edinburgh and the Roslin Institute concluded that rainbow trout exhibit behaviours often associated with pain in other animals. Bee venom and acetic acid injected into the lips resulted in fish rocking their bodies and rubbing their lips along the sides and floors of their tanks, which the researchers concluded were attempts to relieve pain, similar to what mammals would do. Neurons fired in a pattern resembling human neuronal patterns.

Professor James D. Rose of the University of Wyoming claimed the study was flawed since it did not provide proof that fish possess "conscious awareness, particularly a kind of awareness that is meaningfully like ours". Rose argues that since fish brains are so different from human brains, fish are probably not conscious in the manner humans are, so that reactions similar to human reactions to pain instead have other causes. Rose had published a study a year earlier arguing that fish cannot feel pain because their brains lack a neocortex. However, animal behaviourist Temple Grandin argues that fish could still have consciousness without a neocortex because "different species can use different brain structures and systems to handle the same functions."

Animal welfare advocates raise concerns about the possible suffering of fish caused by angling. Some countries, such as Germany have banned specific types of fishing, and the British RSPCA now formally prosecutes individuals who are cruel to fish.

Reproductive Processes

Oogonia development in teleosts fish varies according to the group, and the determination of oogenesis dynamics allows the understanding of maturation and fertilisation processes. Changes in the nucleus, ooplasm, and the surrounding layers characterize the oocyte maturation process.

Postovulatory follicles are structures formed after oocyte release; they do not have endocrine function, present a wide irregular lumen, and are rapidly reabsorbed in a process involving the apoptosis of follicular cells. A degenerative process called follicular atresia reabsorbs vitellogenic oocytes not spawned. This process can also occur, but less frequently, in oocytes in other development stages.

Some fish are hermaphrodites, having both testes and ovaries either at different phases in their life cycle or, as in hamlets, have them simultaneously.

Over 97% of all known fish are oviparous, that is, the eggs develop outside the mother's body. Examples of oviparous fish include salmon, goldfish, cichlids, tuna, and eels. In the majority of these species, fertilisation takes place outside the mother's body, with the male and female fish shedding their gametes into the surrounding water. However, a few oviparous fish practice internal fertilisation, with the male using some sort of intromittent organ to deliver sperm into the genital opening of the female, most notably the oviparous sharks, such as the horn shark, and oviparous rays, such as skates. In these cases, the male is equipped with a pair of modified pelvic fins known as claspers.

Marine fish can produce high numbers of eggs which are often released into the open water column. The eggs have an average diameter of 1 millimetre (0.039 in). The eggs are generally surrounded by the extraembryonic membranes but do not develop a shell, hard or soft, around these membranes. Some fish have thick, leathery coats, especially if they must withstand physical force or desiccation. These type of eggs can also be very small and fragile.

The newly hatched young of oviparous fish are called larvae. They are usually poorly formed, carry a large yolk sac (for nourishment) and are very different in appearance from juvenile and adult specimens. The larval period in oviparous fish is relatively short (usually only several weeks), and larvae rapidly grow and change appearance and structure (a process termed metamorphosis) to become juveniles. During this transition larvae must switch from their yolk sac to feeding on zooplankton prey, a process which depends on typically inadequate zooplankton density, starving many larvae.

In ovoviviparous fish the eggs develop inside the mother's body after internal fertilisation but receive little or no nourishment directly from the mother, depending instead on the yolk. Each embryo develops in its own egg. Familiar examples of ovoviviparous fish include guppies, angel sharks, and coelacanths.

Some species of fish are viviparous. In such species the mother retains the eggs and nourishes the embryos. Typically, viviparous fish have a structure analogous to the placenta seen in mammals connecting the mother's blood supply with that of the embryo.

Examples of viviparous fish include the surf-perches, splitfins, and lemon shark. Some viviparous fish exhibit oophagy, in which the developing embryos eat other eggs produced by the mother. This has been observed primarily among sharks, such as the shortfin mako and porbeagle, but is known for a few bony fish as well, such as the halfbeak *Nomorhamphus ebrardtii*. Intrauterine cannibalism is an even more unusual mode of vivipary, in which the largest embryos eat weaker and smaller siblings. This behaviour is also most commonly found among sharks, such as the grey nurse shark, but has also been reported for *Nomorhamphus ebrardtii*.

In many species of fish, fins have been modified to allow Internal fertilisation.

Aquarists commonly refer to ovoviviparous and viviparous fish as livebearers.

- Many fish species are hermaphrodites. *Synchronous hermaphrodites* possess both ovaries and testes at the same time. *Sequential hermaphrodites* have both types of tissue in their gonads, with one type being predominant while the fish belongs to the corresponding gender.

Chapter 7

Fish Anatomy

Fish anatomy is the study of the form or morphology of fishes. It can be contrasted with fish physiology, which is the study of how the component parts of fish function together in the living fish. In practice, fish anatomy and fish physiology complement each other, the former dealing with the structure of a fish, its organs or component parts and how they are put together, such as might be observed on the dissecting table or under the microscope, and the latter dealing with how those components function together in the living fish.

The anatomy of fish is often shaped by the physical characteristics of water, the medium in which fish live. Water is much denser than air, holds a relatively small amount of dissolved oxygen, and absorbs more light than air does. The body of a fish is divided into a head, trunk and tail, although the divisions between the three are not always externally visible. The skeleton, which forms the support structure inside the fish, is either made of cartilage, in cartilaginous fish, or bone in bony fish.

The main skeletal element is the vertebral column, composed of articulating vertebrae which are lightweight yet strong. The ribs attach to the spine and there are no limbs or limb girdles. The main external features of the fish, the fins, are composed of either bony or soft spines called rays, which with the exception of the caudal fins, have no direct connection with the spine. They are supported by the muscles which compose the main part of the trunk. The heart has two chambers and pumps the blood through the respiratory surfaces of the gills and on round the body in a single circulatory loop. The eyes are adapted for seeing underwater and have only local vision. There is an inner ear but no external or middle ear. Low frequency vibrations are detected by the lateral line system of sense organs that run along the length of the sides of fish, and these respond to nearby movements and to changes in water pressure.

Sharks and rays are basal fish with numerous primitive anatomical features similar to those of ancient fish, including skeletons composed of cartilage. Their bodies tend to be dorso-ventrally flattened, they usually have five pairs of gill slits and a large mouth set on the underside of the head. The dermis is covered with separate dermal placoid scales. They have a cloaca into which the urinary and genital passages open, but not a swim bladder. Cartilaginous fish produce a small number of large, yolky eggs. Some species are ovoviviparous and the young develop internally but others are oviparous and the larvae develop externally in egg cases.

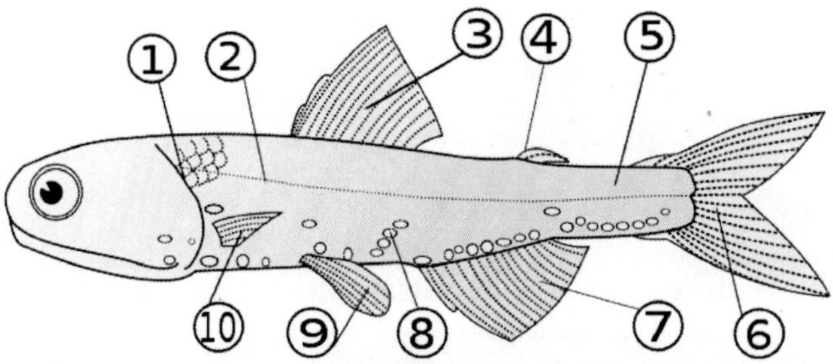

Figure: *External anatomy of a bony fish (Hector's lanternfish) 1. operculum (gill cover), 2. lateral line, 3. dorsal fin, 4. adipose fin, 5. caudal peduncle, 6. caudal fin, 7. anal fin, 8. photophores 9. pelvic fins (paired), 10. pectoral fins (paired)*

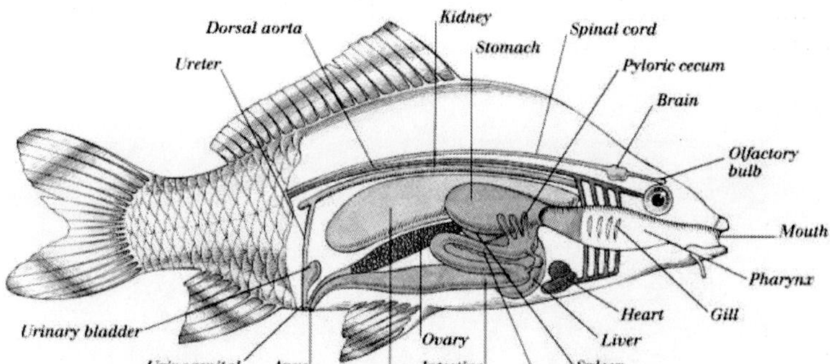

Figure: *Internal anatomy of a bony fish*

The bony fish lineage shows more derived anatomical traits, often with major evolutionary changes from the features of ancient fish. They have a bony skeleton, are generally laterally flattened, have five pairs of gills protected by an operculum, and a mouth at or near the tip of the snout. The dermis is covered with overlapping scales. Bony fish have a swim bladder which helps them maintain a constant depth in the water column, but not a cloaca. They mostly spawn a large number of small eggs with little yolk which they broadcast into the water column.

Body

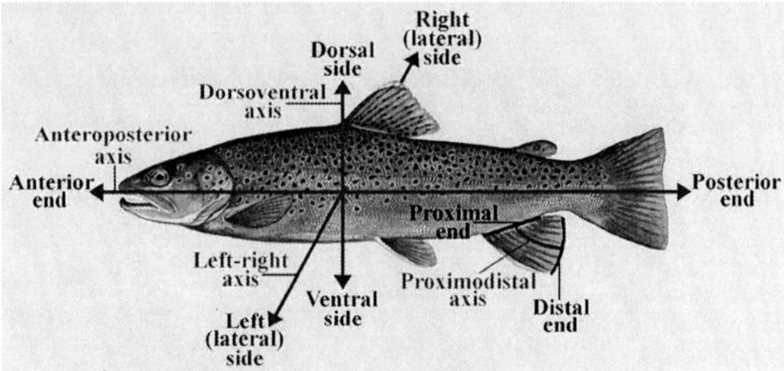

Figure: *Anatomical directions and axes*

In many respects fish anatomy is different from humans and other mammals, yet it shares the same basic vertebrate body plan from which all vertebrates have evolved: a notochord, rudimentary vertebrae, and a well-defined head and tail.

Fish have a variety of different body plans. Their body is divided into head, trunk, and tail, although the divisions are not always externally visible. The body is often fusiform, a streamlined body plan often found in fast-moving fish. They may also be filiform (eel-shaped) or vermiform (worm-shaped). Also, fish are often either compressed (laterally thin) or depressed (dorso-ventrally flat).

Skeleton

Figure: *Skeleton of a bony fish*

There are two different skeletal types: the exoskeleton, which is the stable outer shell of an organism, and the endoskeleton, which forms the support structure inside the body. The skeleton of the fish is either made of cartilage (cartilaginous fishes) or bones (bony fishes). The main features of the fish, the fins, are bony fin rays and, except the caudal fin, have no direct connection with the spine. They are supported only by the muscles. The ribs attach to the spine.

Bones are rigid organs that form part of the endoskeleton of vertebrates. They function to move, support, and protect the various organs of the body, produce red and white blood cells and store minerals. Bone tissue is a type of dense connective tissue. Because bones come in a variety of shapes and have a complex internal and external structure they are lightweight, yet strong and hard, in addition to fulfilling their many other functions.

Fish bones have been used to bioremediate lead from contaminated soil.

Vertebrae

Fish are vertebrates. All vertebrates are built along the basic chordate body plan: a stiff rod running through the length of the animal (vertebral column or notochord), with a hollow tube of nervous tissue (the spinal cord) above it and the gastrointestinal tract below. In all vertebrates, the mouth is found at, or right below, the anterior end of the animal, while the anus opens to the exterior before the end of the body. The remaining part of the body continuing aft of the anus forms a tail with vertebrae and spinal cord, but no gut.

Figure: *The X-ray tetra (Pristella maxillaris) has a visible backbone. The spinal cord is housed within its backbone*

Figure: *Skeleton of a butterfly fish showing the vertebral column*

The defining characteristic of a vertebrate is the vertebral column, in which the notochord (a stiff rod of uniform composition) found in all chordates has been replaced by a segmented series of stiffer elements (vertebrae) separated by mobile joints (intervertebral discs, derived embryonically and evolutionarily from the notochord). However, a few fish have secondarily lost this anatomy, retaining the notochord into adulthood, such as the sturgeon.

The vertebral column consists of a centrum (the central body or spine of the vertebra), vertebral arches which protrude from the top and bottom of the centrum, and various processes which project from the centrum or arches. An arch extending from the top of the centrum is called a neural arch, while the hemal arch or chevron is found underneath the centrum in the caudal (tail) vertebrae of fish. The centrum of a fish is usually concave at each end (amphicoelous), which limits the motion of the fish. This can be contrasted with the centrum of a mammal, which is flat at each end (acoelous), shaped in a manner that can support and distribute compressive forces.

The vertebrae of lobe-finned fishes consist of three discrete bony elements. The vertebral arch surrounds the spinal cord, and is of broadly similar form to that found in most other vertebrates. Just beneath the arch lies a small plate-like *pleurocentrum,* which protects the upper surface of the notochord, and below that, a larger arch-shaped *intercentrum* to protect the lower border. Both of these structures are embedded within a single cylindrical mass of cartilage. A similar arrangement was found in primitive tetrapods, but, in the evolutionary line that led to reptiles (and hence, also to mammals and birds), the intercentrum became partially or wholly replaced by an enlarged pleurocentrum, which in turn became the bony vertebral body.

In most ray-finned fishes, including all teleosts, these two structures are fused with, and embedded within, a solid piece of bone superficially resembling the vertebral body of mammals. In living amphibians, there is simply a cylindrical piece of bone below the vertebral arch, with no trace of the separate elements present in the early tetrapods.

In cartilagenous fish, such as sharks, the vertebrae consist of two cartilagenous tubes. The upper tube is formed from the vertebral arches, but also includes additional cartilagenous structures filling in the gaps between the vertebrae, and so enclosing the spinal cord in an essentially continuous sheath. The lower tube surrounds the notochord, and has a complex structure, often including multiple layers of calcification.

Lampreys have vertebral arches, but nothing resembling the vertebral bodies found in all higher vertebrates. Even the arches are discontinuous, consisting of separate pieces of arch-shaped cartilage around the spinal cord in most parts of the body, changing to long strips of cartilage above and below in the tail region.

Hagfishes lack a true vertebral column, and are therefore not properly considered vertebrates, but a few tiny neural arches are present in the tail. Hagfishes do, however, possess a cranium. For this reason, the vertebrate subphylum is sometimes referred to as "Craniata" when discussing morphology. Molecular analysis since 1992 has suggested that the hagfishes are most closely related to lampreys, and so also are vertebrates in a monophyletic sense. Others consider them a sister group of vertebrates in the common taxon of Craniata.

Head

The head includes the snout, from the eye to the forward most point of the upper jaw, the operculum or gill cover (absent in sharks and jawless fish), and the cheek, which extends from the eye to preopercle. The operculum and preopercle may or may not have spines. In sharks and some primitive bony fish a spiracle, small extra gill opening, is found behind each eye.

The skull in fishes is formed from a series of only loosely connected bones. Jawless fish and sharks only possess a cartilaginous endocranium, with the upper and lower jaws being separate elements.

Bony fishes have additional dermal bone, forming a more or less coherent skull roof in lungfish and holost fish. The lower jaw defines a chin.

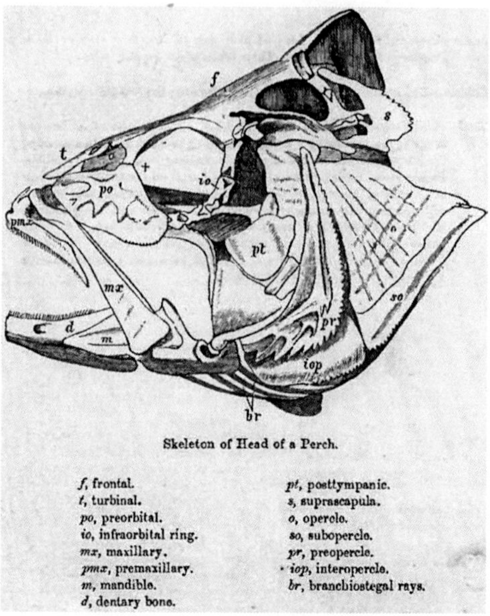

***Figure:** Parts of the head*

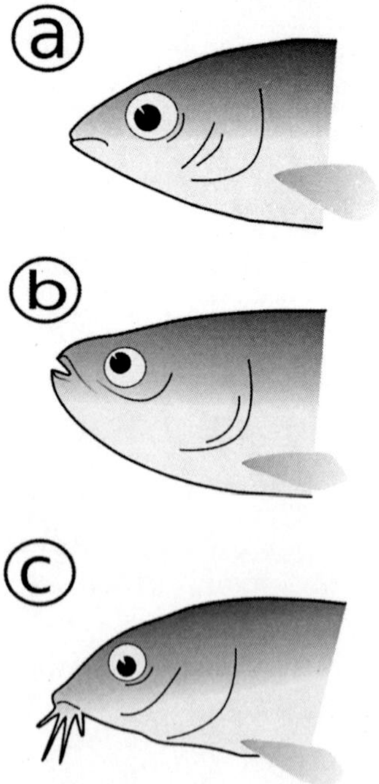

***Figure:** Positions of fish mouths: (a) - terminal, (b) - superior, (c) - subterminal, inferior*

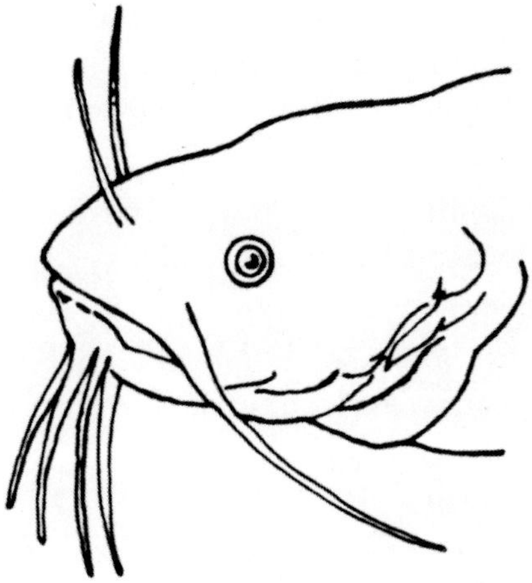

Figure: *Barbels*

In lampreys, the mouth is formed into an oral disk. In most jawed fish, however, there are three general configurations. The mouth may be on the forward end of the head (*terminal*), may be upturned (*superior*), or may be turned downwards or on the bottom of the fish (*subterminal* or *inferior*). The mouth may be modified into a suckermouth adapted for clinging onto objects in fast-moving water.

The simpler structure is found in jawless fish, in which the cranium is represented by a trough-like basket of cartilaginous elements only partially enclosing the brain, and associated with the capsules for the inner ears and the single nostril. Distinctively, these fish have no jaws.

Cartilaginous fish, such as sharks, have also simple, and presumably primitive, skull structures. The cranium is a single structure forming a case around the brain, enclosing the lower surface and the sides, but always at least partially open at the top as a large fontanelle. The most anterior part of the cranium includes a forward plate of cartilage, the rostrum, and capsules to enclose the olfactory organs. Behind these are the orbits, and then an additional pair of capsules enclosing the structure of the inner ear. Finally, the skull tapers towards the rear, where the foramen magnum lies immediately above a single condyle, articulating with the first vertebra. There are, in addition, at various points throughout the cranium, smaller foramina for the cranial nerves. The jaws consist of separate hoops of cartilage, almost always distinct from the cranium proper.

In the ray-finned fishes, there has also been considerable modification from the primitive pattern. The roof of the skull is generally well formed, and although the exact relationship of its bones to those of tetrapods is unclear, they are usually given similar names for convenience. Other elements of the skull, however, may be reduced; there is little cheek region behind the enlarged orbits, and little, if any bone in between them.

The upper jaw is often formed largely from the premaxilla, with the maxilla itself located further back, and an additional bone, the symplectic, linking the jaw to the rest of the cranium.

Although the skulls of fossil lobe-finned fish resemble those of the early tetrapods, the same cannot be said of those of the living lungfishes. The skull roof is not fully formed, and consists of multiple, somewhat irregularly shaped bones with no direct relationship to those of tetrapods. The upper jaw is formed from the pterygoids and vomers alone, all of which bear teeth. Much of the skull is formed from cartilage, and its overall structure is reduced.

The head may have several fleshy structures known as barbels, which may be very long and resemble whiskers. Many fish species also have a variety of protrusions or spines on the head. The nostrils or nares of almost all fishes do not connect to the oral cavity, but are pits of varying shape and depth.

External Organs

Jaw

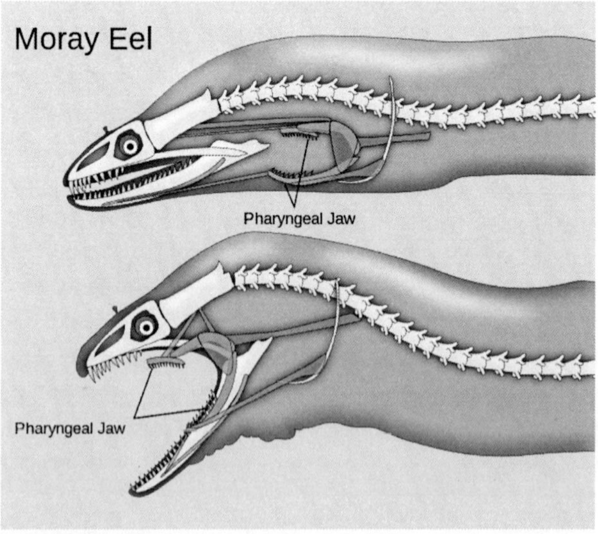

Figure:.*Moray eels have two sets of jaws: the oral jaws that capture prey and the pharyngeal jaws that advance into the mouth and move prey from the oral jaws to the esophagus for swallowing*

The vertebrate jaw probably originally evolved in the Silurian period and appeared in the Placoderm fish which further diversified in the Devonian. Jaws are thought to derive from the pharyngeal arches that support the gills in fish.

The two most anterior of these arches are thought to have become the jaw itself and the hyoid arch, which braces the jaw against the braincase and increases mechanical efficiency. While there is no fossil evidence directly to support this theory, it makes sense in light of the numbers of pharyngeal arches that are visible in extant jawed (the Gnathostomes), which have seven arches, and primitive jawless vertebrates (the Agnatha), which have nine.

It is thought that the original selective advantage garnered by the jaw was not related to feeding, but to increased respiration efficiency. The jaws were used in the buccal pump (observable in modern fish and amphibians) that pumps water across the gills of fish or air into the lungs in the case of amphibians. Over evolutionary time the more familiar use of jaws (to humans), in feeding, was selected for and became a very important function in vertebrates.

Linkage systems are widely distributed in animals. The most thorough overview of the different types of linkages in animals has been provided by M. Muller, who also designed a new classification system, which is especially well suited for biological systems. Linkage mechanisms are especially frequent and manifold in the head of bony fishes, such as wrasses, which have evolved many specialized feeding mechanisms. Especially advanced are the linkage mechanisms of jaw protrusion. For suction feeding a system of linked four-bar linkages is responsible for the coordinated opening of the mouth and 3-D expansion of the buccal cavity. Other linkages are responsible for protrusion of the premaxilla.

Eyes

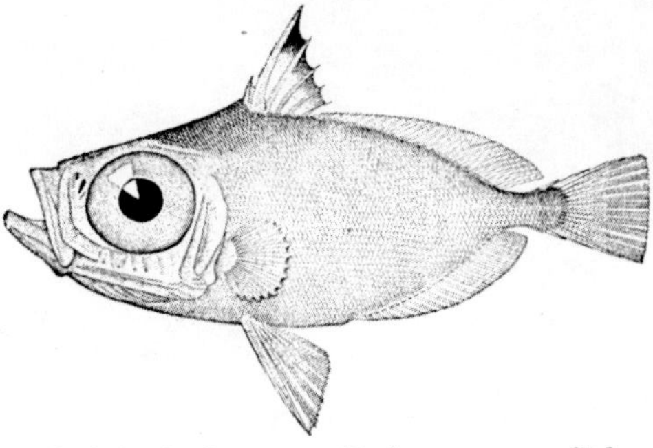

Figure: *Zenion hololepis is a small deep water fish with large eyes*

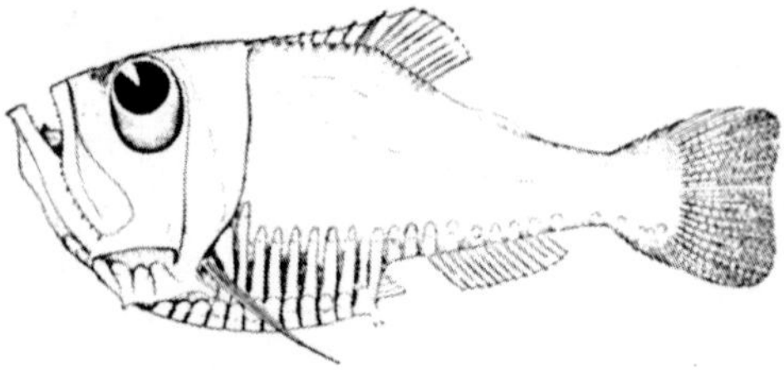

Figure: *The deep sea half-naked hatchetfish has eyes which look overhead where it can see the silhouettes of prey*

Fish eyes are similar to terrestrial vertebrates like birds and mammals, but have a more spherical lens. Their retinas generally have both rod cells and cone cells (for scotopic and photopic vision), and most species have colour vision. Some fish can see ultraviolet and some can see polarized light. Amongst jawless fish, the lamprey has well-developed eyes, while the hagfish has only primitive eyespots. The ancestors of modern hagfish, thought to be the protovertebrate were evidently pushed to very deep, dark waters, where they were less vulnerable to sighted predators, and where it is advantageous to have a convex eye-spot, which gathers more light than a flat or concave one. Unlike humans, fish normally adjust focus by moving the lens closer to or further from the retina.

Gills

Figure: *Gill of a rainbow trout*

The gills, located under the operculum, are a respiratory organ for the extraction of oxygen from water and for the excretion of carbon dioxide. They are not usually visible, but can be seen in some species, such as the frilled shark. The labyrinth organ of Anabantoidei and Clariidae is used to allow the fish to extract oxygen from the air. Gill rakers are bony or cartilaginous, finger-like projections off the gill arch which function in filter-feeders to retain filtered prey.

Skin

The epidermis of fish consists entirely of live cells, with only minimal quantities of keratin in the cells of the superficial layer. It is generally permeable. The dermis of bony fish typically contains relatively little of the connective tissue found in tetrapods.

Instead, in most species, it is largely replaced by solid, protective bony scales. Apart from some particularly large dermal bones that form parts of the skull, these scales are lost in tetrapods, although many reptiles do have scales of a different kind, as do pangolins. Cartilaginous fish have numerous tooth-like denticles embedded in their skin, in place of true scales.

Sweat glands and sebaceous glands are both unique to mammals, but other types of skin glands are found in fish. Fish typically have numerous individual mucus-secreting skin cells that aid in insulation and protection, but may also have poison glands, photophores, or cells that produce a more watery, serous fluid. Melanin colours the skin of many species, but in fish the epidermis is often relatively colourless. Instead, the colour of the skin is largely due to chromatophores in the dermis, which, in addition to melanin, may contain guanine or carotenoid pigments. Many species, such as flounders, change the colour of their skin by adjusting the relative size of their chromatophores.

Scales

Figure: *Cycloid scales covering rohu*

The outer body of many fish is covered with scales, which are part of the fish's integumentary system. The scales originate from the mesoderm (skin), and may be similar in structure to teeth. Some species are covered instead by scutes. Others have no outer covering on the skin. Most fish are covered in a protective layer of slime (mucus).

There are four principal types of fish scales.

1. Placoid scales, also called dermal denticles, are similar to teeth in that they are made of dentin covered by enamel. They are typical of sharks and rays.
2. Ganoid scales are flat, basal-looking scales that cover a fish body with little overlapping. They are typical of gar and bichirs.
3. Cycloid scales are small oval-shaped scales with growth rings. Bowfin and remora have cycloid scales.
4. Ctenoid scales are similar to the cycloid scales, with growth rings. They are distinguished by spines that cover one edge. Halibut have this type of scale.

Another, less common, type of scale is the scute, which is:

- an external shield-like bony plate, or
- a modified, thickened scale that often is keeled or spiny, or
- a projecting, modified (rough and strongly ridged) scale, usually associated with the lateral line, or on the caudal peduncle forming caudal keels, or along the ventral profile. Some fish, such as pineconefish, are completely or partially covered in scutes.

Lateral Line

Figure: *The lateral line is clearly visible as a line of receptors running along the side of this Atlantic cod*

The lateral line is a sense organ used to detect movement and vibration in the surrounding water. For example, fish can use their lateral line system to follow the vortices produced by fleeing prey. In most species, it consists of a line of receptors running along each side of the fish.

Photophores

Photophores are light-emitting organs which appears as luminous spots on some fishes. The light can be produced from compounds during the digestion of prey, from specialized mitochondrial cells in the organism called photocytes, or associated with symbiotic bacteria, and are used for attracting food or confusing predators.

Fins

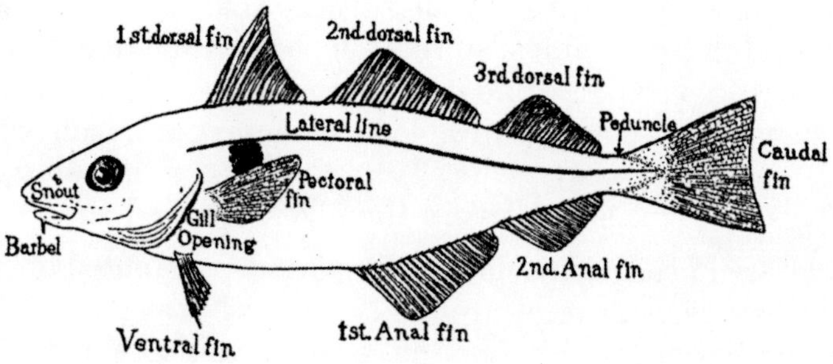

Figure: *The haddock, a type of cod, is ray-finned. It has three dorsal and two anal fins*

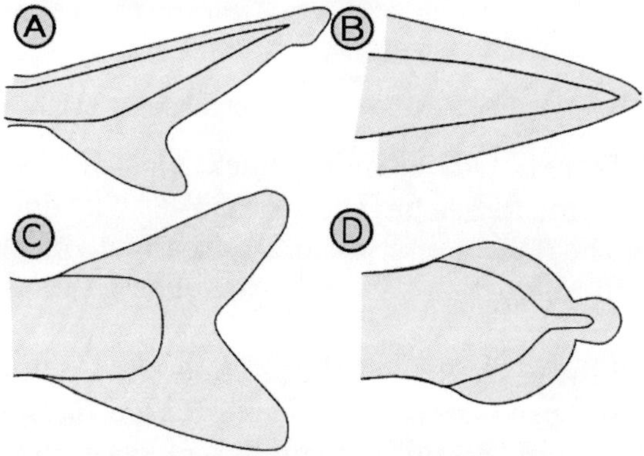

Figure: *Types of caudal (tail) fin: (A) - Heterocercal, (B) - Protocercal, (C) - Homocercal, (D) - Diphycercal*

Fins are the most distinctive features of fish. They are either composed of bony spines protruding from the body with skin covering them and joining them together, either in a webbed fashion as seen in most bony fish, or are similar to a flipper, as seen in sharks. Apart from the tail or caudal fin, fins have no direct connection with the spine and are supported by muscles only. Their principal function is to help the fish swim. Fins can also be used for gliding or crawling, as seen in the flying fish and frogfish. Fins located in different places on the fish serve different purposes, such as moving forward, turning, and keeping an upright position. For every fin, there are a number of fish species in which this particular fin has been lost during evolution.

Spines and Rays

In bony fish, most fins may have spines or rays. A fin may contain only spiny rays, only soft rays, or a combination of both. If both are present, the spiny rays are always anterior. Spines are generally stiff and sharp. Rays are generally soft, flexible, segmented, and may be branched. This segmentation of rays is the main difference that separates them from spines; spines may be flexible in certain species, but they will never be segmented. Spines have a variety of uses. In catfish, they are used as a form of defence; many catfish have the ability to lock their spines outwards. Triggerfish also use spines to lock themselves in crevices to prevent them being pulled out.

Lepidotrichia are bony, bilaterally-paired, segmented fin rays found in bony fishes. They develop around actinotrichia as part of the dermal exoskeleton. Lepidotrichia may have some cartilage or bone in them as well. They are actually segmented and appear as a series of disks stacked one on top of another. The genetic basis for the formation of the fin rays is thought to be genes coding for the proteins actinodin 1 and actinodin 2.

Types of Fin

- Dorsal fins are located on the back. Most fishes have one dorsal fin, but some fishes have two or three . The dorsal fins serve to protect the fish against rolling, and assists in sudden turns and stops. In anglerfish, the anterior of the dorsal fin is modified into an *illicium* and *esca*, a biological equivalent to a fishing rod and lure. The bones that support the dorsal fin are called *Pterygiophore*. There are two to three of them: "proximal", "middle", and "distal". In spinous fins the distal is often fused to the middle, or not present at all.
- The caudal fin is the tail fin, located at the end of the caudal peduncle and is used for propulsion. The caudal peduncle is the narrow part of the fish's body to which the caudal or tail fin is attached. The hypural joint is the joint between the caudal fin and the last of the vertebrae. The hypural is often fan-shaped. The tail is called:
    - o *Heterocercal* if the vertebrae extend into the upper lobe of the tail, making it longer (as in sharks)
    - o *Reversed heterocercal* if the vertebrae extend into the lower lobe of the tail, making it longer (as in the Anaspida)
    - o *Protocercal* if the vertebrae extend to the tip of the tail and the tail is symmetrical but not expanded (as in amphioxus)
    - o *Diphycercal* if the vertebrae extend to the tip of the tail and the tail is symmetrical and expanded (as in the bichir, lungfish, lamprey and coelacanth. Most Palaeozoic fishes had a diphycercal heterocercal tail.)

- o Most fish have a *homocercal* tail, where the fin appears superficially symmetric but the vertebrae extend for a very short distance into the upper lobe of the fin. This can be expressed in a variety of shapes. The tail fin can be:
  - — rounded at the end
  - — truncated: or end in a more-or-less vertical edge, such as in salmon
  - — forked: or end in two prongs
  - — emarginate: or with a slight inward curve.
  - — continuous: with dorsal, caudal and anal fins attached, such as in eels

- The anal fin is located on the ventral surface behind the anus. This fin is used to stabilize the fish while swimming.
- The paired pectoral fins are located on each side, usually just behind the operculum, and are homologous to the forelimbs of tetrapods. A peculiar function of pectoral fins, highly developed in some fish, is the creation of the dynamic lifting force that assists some fish, such as sharks, in maintaining depth and also enables the "flight" for flying fish. In many fish, the pectoral fins aid in walking, especially in the lobe-like fins of some anglerfish and in the mudskipper. Certain rays of the pectoral fins may be adapted into finger-like projections, such as in sea robins and flying gurnards. The "horns" of manta rays and their relatives are called *cephalic fins*; this is actually a modification of the anterior portion of the pectoral fin.
- The paired pelvic or ventral fins are located ventrally below the pectoral fins. They are homologous to the hindlimbs of tetrapods. The pelvic fin assists the fish in going up or down through the water, turning sharply, and stopping quickly. In gobies, the pelvic fins are often fused into a single sucker disk. This can be used to attach to objects.
- The adipose fin is a soft, fleshy fin found on the back behind the dorsal fin and just forward of the caudal fin. It is absent in many fish families, but is found in Salmonidae, characins and catfishes. Its function has remained a mystery, and is frequently clipped off to mark hatchery-raised fish, though data from 2005 showed that trout with their adipose fin removed have an 8% higher tailbeat frequency. Additional research published in 2011 has suggested that the fin may be vital for the detection of and response to stimuli such as touch, sound and changes in

pressure. Canadian researchers identified a neural network in the fin, indicating that it likely has a sensory function, but are still not sure exactly what the consequences of removing it are.

- Some types of fast-swimming fish have a horizontal caudal keel just forward of the tail fin. Much like the keel of a ship, this is a lateral ridge on the caudal peduncle, usually composed of scutes, that provides stability and support to the caudal fin. There may be a single paired keel, one on each side, or two pairs above and below.
- Finlets are small fins, generally between the dorsal and the caudal fins also between the anal fin and the caudal fin (in bichirs, there are only finlets on the dorsal surface and no dorsal fin). In some fish such as tuna or sauries, they are rayless, non-retractable, and found between the last dorsal and/or anal fin and the caudal fin.

Internal Organs

Intestines

As with other vertebrates, the intestines of fish consist of two segments, the small intestine and the large intestine. In most higher vertebrates, the small intestine is further divided into the duodenum and other parts. In fish, the divisions of the small intestine are not as clear, and the terms *anterior intestine* or *proximal intestine* may be used instead of duodenum. In bony fish, the intestine is relatively short, typically around one and a half times the length of the fish's body. It commonly has a number of *pyloric caeca*, small pouch-like structures along its length that help to increase the overall surface area of the organ for digesting food.

There is no ileocaecal valve in teleosts, with the boundary between the small intestine and the rectum being marked only by the end of the digestive epithelium. There is no small intestine as such in non-teleost fish, such as sharks, sturgeons, and lungfish. Instead, the digestive part of the gut forms a *spiral intestine*, connecting the stomach to the rectum. In this type of gut, the intestine itself is relatively straight, but has a long fold running along the inner surface in a spiral fashion, sometimes for dozens of turns.

This fold creates a valve-like structure that greatly increases both the surface area and the effective length of the intestine. The lining of the spiral intestine is similar to that of the small intestine in teleosts and non-mammalian tetrapods. In lampreys, the spiral valve is extremely small, possibly because their diet requires little digestion. Hagfish have no spiral valve at all, with digestion occurring for almost the entire length of the intestine, which is not subdivided into different regions.

Pyloric Caeca

The pyloric caecum is a pouch, usually peritoneal, at the beginning of the large intestine. It receives faecal material from the ileum, and connects to the ascending colon of the large intestine. It is present in most amniotes, and also in lungfish. Many fish in addition have a number of small outpocketings, also called pyloric caeca, along their intestine; despite the name they are not homologous with the caecum of amniotes, and their purpose is to increase the overall area of the digestive epithelium.

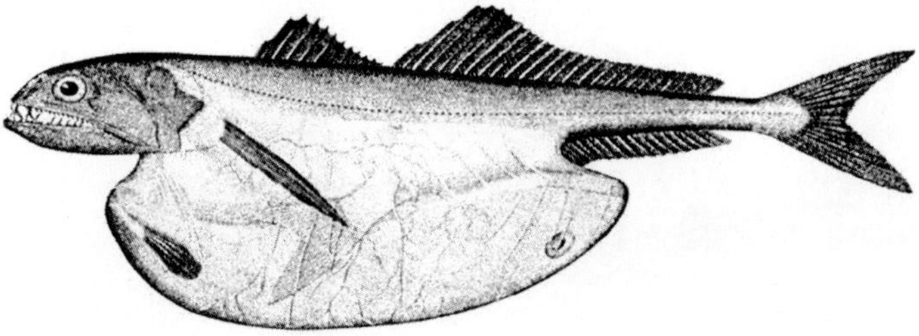

***Figure:** The black swallower is a species of deep sea fish with an extensible stomach which allows it to swallow fish larger than itself*

***Figure:** Internal organs of a female Atlantic cod 1. Liver, 2. Gas bladder, 3. ovary, 4. Pyloric caeca, 5. Stomach, 6. Intestine*

Stomach

As with other vertebrates, the relative positions of the esophageal and duodenal openings to the stomach remain relatively constant. As a result, the stomach always curves somewhat to the left before curving back to meet the pyloric sphincter. However, lampreys, hagfishes, chimaeras, lungfishes, and some teleost fish have no stomach at all, with the esophagus opening directly into the intestine. These fish consume diets that either require little storage of food, or no pre-digestion with gastric juices, or both.

Kidneys

The kidneys of fish are typically narrow, elongated organs, occupying a significant portion of the trunk. They are similar to the mesonephros of higher vertebrates (reptiles, birds and mammals). The kidneys contain clusters of nephrons, serviced by collecting ducts which usually drain into a mesonephric duct. However, the situation is not always so simple. In cartilaginous fish there is also a shorter duct which drains the posterior (metanephric) parts of the kidney, and joins with the mesonephric duct at the bladder or cloaca. Indeed, in many cartilaginous fish, the anterior portion of the kidney may degenerate or cease to function altogether in the adult. Hagfish and lamprey kidneys are unusually simple. They consist of a row of nephrons, each emptying directly into the mesonephric duct.

Spleen

The spleen is found in nearly all vertebrates. It is a non-vital organ, similar in structure to a large lymph node. It acts primarily as a blood filter, and plays important roles in regard to red blood cells and the immune system. In cartilaginous and bony fish it consists primarily of red pulp and is normally a somewhat elongated organ as it actually lies inside the serosal lining of the intestine. The only vertebrates lacking a spleen are the lampreys and hagfishes. Even in these animals, there is a diffuse layer of haematopoeitic tissue within the gut wall, which has a similar structure to red pulp, and is presumed to be homologous with the spleen of higher vertebrates.

Liver

The liver is a large vital organ present in all fish. It has a wide range of functions, including detoxification, protein synthesis, and production of biochemicals necessary for digestion.

Heart

Fish have what is often described as a two-chambered heart, consisting of one atrium to receive blood and one ventricle and pump it, in contrast to three chambers (two atria, one ventricle) of amphibian and most reptile hearts and four chambers (two atria, two ventricles) of mammal and bird hearts.

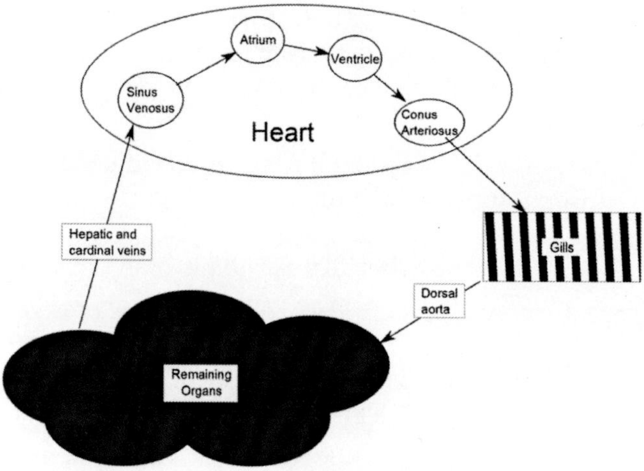

Figure: *Cardiovascular cycle in a fish*

However, the fish heart has entry and exit compartments that may be called chambers, so it is also sometimes described as three-chambered or four-chambered, depending on what is counted as a chamber. The atrium and ventricle are sometimes considered "true chambers", while the others are considered "accessory chambers".

The four compartments are arranged sequentially:

- Sinus venosus, a thin-walled sac or reservoir with some cardiac muscle that collects deoxygenated blood through the incoming hepatic and cardinal veins.
- Atrium, a thicker-walled, muscular chamber that sends blood to the ventricle.
- Ventricle, a thick-walled, muscular chamber that pumps the blood to the fourth part, the outflow tract. The shape of the ventricle varies considerably, usually tubular in fish with elongated bodies, pyramidal with a triangular base in others, or sometimes sac-like in some marine fish.
- The outflow tract (OFT) to the ventral aorta, consisting of the tubular conus arteriosus, bulbus arteriosus, or both. The conus arteriosus, typically found in more primitive species of fish, contracts to assist blood flow to the aorta, while the bulbus anteriosus does not.

Ostial valves, consisting of flap-like connective tissues, prevent blood from flowing backward through the compartments. The ostial valve between the sinus venosus and atrium is called the sino-atrial valve, which closes during ventricular contraction. Between the atrium and ventricle is an ostial valve called the atrio-ventricular valve, and between the bulbus arteriosus and ventricle is an ostial valve called the bulbo-ventricular valve. The conus arteriosus has a variable number of semilunar valves.

The ventral aorta delivers blood to the gills where it is oxygenated and flows, through the dorsal aorta, into the rest of the body. (In tetrapods, the ventral aorta has divided in two; one half forms the ascending aorta, while the other forms the pulmonary artery).

The circulatory systems of all vertebrates, are *closed*. Fish have the simplest circulatory system, consisting of only one circuit, with the blood being pumped through the capillaries of the gills and on to the capillaries of the body tissues. This is known as *single cycle* circulation.

In the adult fish, the four compartments are not arranged in a straight row but, instead form an S-shape with the latter two compartments lying above the former two. This relatively simpler pattern is found in cartilaginous fish and in the ray-finned fish. In teleosts, the conus arteriosus is very small and can more accurately be described as part of the aorta rather than of the heart proper. The conus arteriosus is not present in any amniotes, presumably having been absorbed into the ventricles over the course of evolution. Similarly, while the sinus venosus is present as a vestigial structure in some reptiles and birds, it is otherwise absorbed into the right atrium and is no longer distinguishable.

Swim Bladder

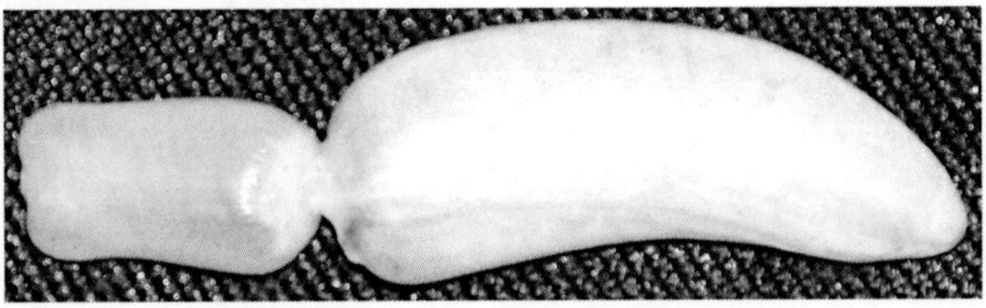

Figure: *The swim bladder of a rudd*

The swim bladder (or gas bladder) is an internal organ that contributes to the ability of a fish to control its buoyancy, and thus to stay at the current water depth, ascend, or descend without having to waste energy in swimming. The bladder is only found in the bony fishes. In the more primitive groups like some minnows, bichirs and lungfish, the bladder is open to the esophagus and double as a lung. It is often absent in fast swimming fishes such as the tuna and mackerel families. The condition of a bladder open to the esophagus is called physostome, the closed condition physoclist. In the latter, the gas content of the bladder is controlled through a rete mirabilis, a network of blood vessels effecting gas exchange between the bladder and the blood.

Weberian Apparatus

Fishes of the superorder Ostariophysi possess a structure called the Weberian apparatus, a modification which allow them to hear better. This ability which may well explain the marked success of otophysian fishes. The apparatus is made up of a set of bones known as *Weberian ossicles*, a chain of small bones that connect the auditory system to the swim bladder of fishes. The ossicles connect the gas bladder wall with Y-shaped lymph sinus that abuts the lymph-filled transverse canal joining the sacculi of the right and left ears. This allows the transmission of vibrations to the inner ear. A fully functioning Weberian apparatus consists of the swim bladder, the Weberian ossicles, a portion of the anterior vertebral column, and some muscles and ligaments.

Reproductive Organs

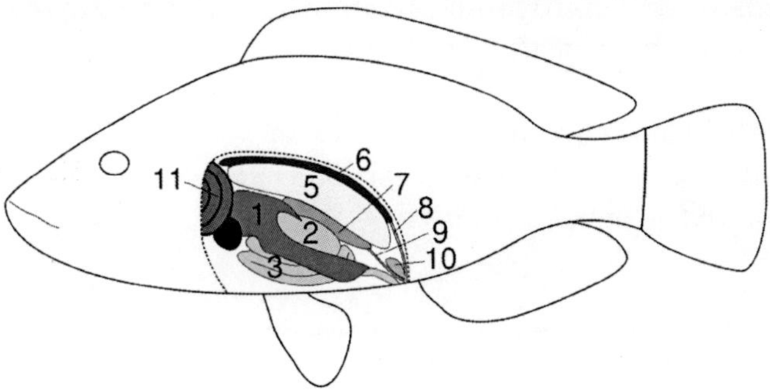

Figure: *7 = testicles or ovaries*

Fish reproductive organs include testes and ovaries. In most species, gonads are paired organs of similar size, which can be partially or totally fused. There may also be a range of secondary organs that increase reproductive fitness. The genital papilla is a small, fleshy tube behind the anus in some fishes, from which the sperm or eggs are released; the sex of a fish often can be determined by the shape of its papilla.

Testes

Most male fish have two testes of similar size. In the case of sharks, the testes on the right side is usually larger. The primitive jawless fish have only a single testis, located in the midline of the body, although even this forms from the fusion of paired structures in the embryo.

Under a tough membranous shell, the tunica albuginea, the testis of some teleost fish, contains very fine coiled tubes called seminiferous tubules. The tubules are lined with a layer of cells (germ cells) that from puberty into old age, develop into sperm cells (also known as spermatozoa or male gametes). The developing sperm travel through the seminiferous tubules to the rete testis located in the mediastinum testis, to the efferent ducts, and then to the epididymis where newly created sperm cells mature. The sperm move into the vas deferens, and are eventually expelled through the urethra and out of the urethral orifice through muscular contractions.

However, most fish do not possess seminiferous tubules. Instead, the sperm are produced in spherical structures called *sperm ampullae*. These are seasonal structures, releasing their contents during the breeding season, and then being reabsorbed by the body. Before the next breeding season, new sperm ampullae begin to form and ripen. The ampullae are otherwise essentially identical to the seminiferous tubules in higher vertebrates, including the same range of cell types.

In terms of spermatogonia distribution, the structure of teleosts testes has two types: in the most common, spermatogonia occur all along the seminiferous tubules, while in Atherinomorph fish they are confined to the distal portion of these structures. Fish can present cystic or semi-cystic spermatogenesis in relation to the release phase of germ cells in cysts to the seminiferous tubules lumen.

Ovaries

Many of the features found in ovaries are common to all vertebrates, including the presence of follicular cells and tunica albuginea There may be hundreds or even millions of fertile eggs present in the ovary of a fish at any given time. Fresh eggs may be developing from the germinal epithelium throughout life. Corpora lutea are found only in mammals, and in some elasmobranch fish; in other species, the remnants of the follicle are quickly resorbed by the ovary. The ovary of teleosts is often contains a hollow, lymph-filled space which opens into the oviduct, and into which the eggs are shed. Most normal female fish have two ovaries. In some elasmobranchs, only the right ovary develops fully. In the primitive jawless fish, and some teleosts, there is only one ovary, formed by the fusion of the paired organs in the embryo.

Fish ovaries may be of three types: gymnovarian, secondary gymnovarian or cystovarian. In the first type, the oocytes are released directly into the coelomic cavity and then enter the ostium, then through the oviduct and are eliminated.

Secondary gymnovarian ovaries shed ova into the coelom from which they go directly into the oviduct. In the third type, the oocytes are conveyed to the exterior through the oviduct. Gymnovaries are the primitive condition found in lungfish, sturgeon, and bowfin. Cystovaries characterize most teleosts, where the ovary lumen has continuity with the oviduct. Secondary gymnovaries are found in salmonids and a few other teleosts.

Nervous System

Central Nervous System

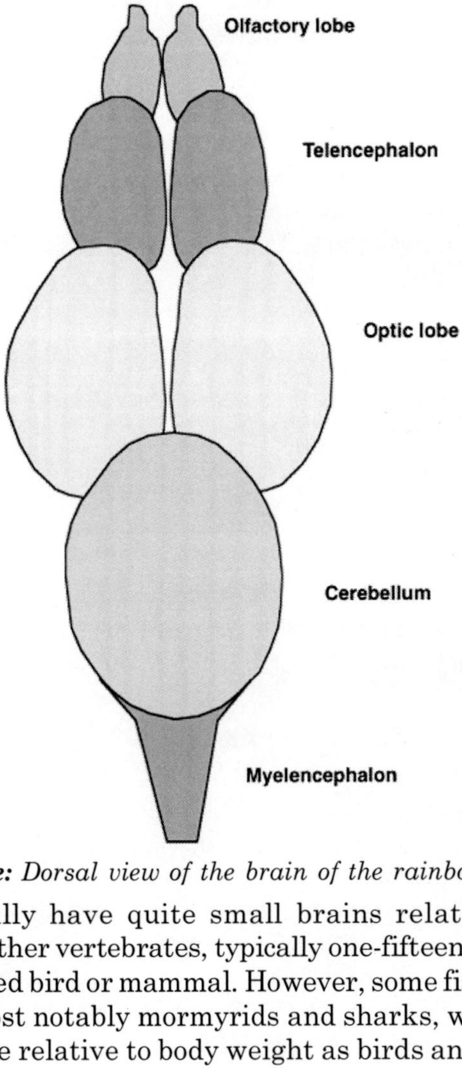

Figure: *Dorsal view of the brain of the rainbow trout*

Fish typically have quite small brains relative to body size compared with other vertebrates, typically one-fifteenth the brain mass of a similarly sized bird or mammal. However, some fish have relatively large brains, most notably mormyrids and sharks, which have brains about as massive relative to body weight as birds and marsupials.

Fish brains are divided into several regions. At the front are the olfactory lobes, a pair of structures that receive and process signals from the nostrils via the two olfactory nerves. The olfactory lobes are very large in fish that hunt primarily by smell, such as hagfish, sharks, and catfish. Behind the olfactory lobes is the two-lobed telencephalon, the structural equivalent to the cerebrum in higher vertebrates. In fish the telencephalon is concerned mostly with olfaction. Together these structures form the forebrain.

The forebrain is connected to the midbrain via the diencephalon. The diencephalon performs functions associated with hormones and homeostasis. The pineal body lies just above the diencephalon. This structure detects light, maintains circadian rhythms, and controls colour changes. The midbrain or mesencephalon contains the two optic lobes. These are very large in species that hunt by sight, such as rainbow trout and cichlids.

The hindbrain or metencephalon is particularly involved in swimming and balance. The cerebellum is a single-lobed structure that is typically the biggest part of the brain. Hagfish and lampreys have relatively small cerebellae, while the mormyrid cerebellum is massive and apparently involved in their electrical sense.

The brain stem or myelencephalon is the brain's posterior. As well as controlling some muscles and body organs, in bony fish at least, the brain stem governs respiration and osmoregulation.

Vertebrates are the only chordate group to exhibit a proper brain. A slight swelling of the anterior end of the dorsal nerve cord is found in the lancelet, though it lacks the eyes and other complex sense organs comparable to those of vertebrates. Other chordates do not show any trends towards cephalisation. The central nervous system is based on a hollow nerve tube running along the length of the animal, from which the peripheral nervous system branches out to enervate the various systems.

The front end of the nerve tube is expanded by a thickening of the walls and expansion of the central canal of spinal cord into three primary brain vesicles: The prosencephalon (forebrain), mesencephalon (midbrain) and rhombencephalon (hindbrain), further differentiated in the various vertebrate groups. Two laterally placed eyes form around outgrows from the midbrain, except in hagfish, though this may be a secondary loss. The forebrain is well developed and subdivided in most tetrapods, while the midbrain dominate in many fish and some salamanders. Vesicles of the forebrain are usually paired, giving rise to hemispheres like the cerebral hemispheres in mammals. The resulting anatomy of the central nervous system, with a single, hollow ventral nerve cord topped by a series of (often paired) vesicles is unique to vertebrates.

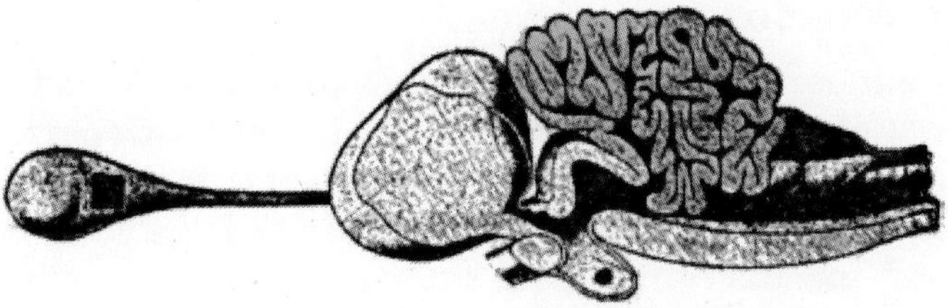

Figure: *Cross-section of the brain of a porbeagle shark, with the cerebellum highlighted*

Cerebellum

The circuits in the cerebellum are similar across all classes of vertebrates, including fish, reptiles, birds, and mammals. There is also an analogous brain structure in cephalopods with well-developed brains, such as octopuses. This has been taken as evidence that the cerebellum performs functions important to all animal species with a brain.

There is considerable variation in the size and shape of the cerebellum in different vertebrate species. In amphibians, lampreys, and hagfish, the cerebellum is little developed; in the latter two groups, it is barely distinguishable from the brain-stem. Although the spinocerebellum is present in these groups, the primary structures are small paired nuclei corresponding to the vestibulocerebellum.

The cerebellum of cartilaginous and bony fishes is extraordinarily large and complex. In at least one important respect, it differs in internal structure from the mammalian cerebellum: The fish cerebellum does not contain discrete deep cerebellar nuclei. Instead, the primary targets of Purkinje cells are a distinct type of cell distributed across the cerebellar cortex, a type not seen in mammals. In mormyrid fish (a family of weakly electrosensitive freshwater fish), the cerebellum is considerably larger than the rest of the brain put together. The largest part of it is a special structure called the *valvula*, which has an unusually regular architecture and receives much of its input from the electrosensory system.

Most species of fish and amphibians possess a lateral line system that senses pressure waves in water. One of the brain areas that receives primary input from the lateral line organ, the medial octavolateral nucleus, has a cerebellum-like structure, with granule cells and parallel fibres. In electrosensitive fish, the input from the electrosensory system goes to the dorsal octavolateral nucleus, which also has a cerebellum-like structure. In ray-finned fishes (by far the largest group), the optic tectum has a layer — the marginal layer — that is cerebellum-like.

Identified Neurons

A neuron is called *identified* if it has properties that distinguish it from every other neuron in the same animal—properties such as location, neurotransmitter, gene expression pattern, and connectivity—and if every individual organism belonging to the same species has one and only one neuron with the same set of properties. In vertebrate nervous systems very few neurons are "identified" in this sense—in humans, there are believed to be none—but in simpler nervous systems, some or all neurons may be thus unique.

In vertebrates, the best known identified neurons are the gigantic Mauthner cells of fish. Every fish has two Mauthner cells, located in the bottom part of the brainstem, one on the left side and one on the right. Each Mauthner cell has an axon that crosses over, innervating neurons at the same brain level and then travelling down through the spinal cord, making numerous connections as it goes. The synapses generated by a Mauthner cell are so powerful that a single action potential gives rise to a major behavioural response: within milliseconds the fish curves its body into a C-shape, then straightens, thereby propelling itself rapidly forward. Functionally this is a fast escape response, triggered most easily by a strong sound wave or pressure wave impinging on the lateral line organ of the fish. Mauthner cells are not the only identified neurons in fish—there are about 20 more types, including pairs of "Mauthner cell analogs" in each spinal segmental nucleus. Although a Mauthner cell is capable of bringing about an escape response all by itself, in the context of ordinary behaviour other types of cells usually contribute to shaping the amplitude and direction of the response.

Mauthner cells have been described as command neurons. A command neuron is a special type of identified neuron, defined as a neuron that is capable of driving a specific behaviour all by itself. Such neurons appear most commonly in the fast escape systems of various species—the squid giant axon and squid giant synapse, used for pioneering experiments in neurophysiology because of their enormous size, both participate in the fast escape circuit of the squid. The concept of a command neuron has, however, become controversial, because of studies showing that some neurons that initially appeared to fit the description were really only capable of evoking a response in a limited set of circumstances.

Immune System

Immune organs vary by type of fish. In the jawless fish (lampreys and hagfish), true lymphoid organs are absent. These fish rely on regions of lymphoid tissue within other organs to produce immune cells. For example, erythrocytes, macrophages and plasma cells are produced in the anterior kidney (or pronephros) and some areas of the gut (where granulocytes mature.) They resemble primitive bone marrow in hagfish. Cartilaginous fish (sharks and rays) have a more advanced immune system. They have three specialized organs that are unique to chondrichthyes; the epigonal organs (lymphoid tissue similar to mammalian bone) that surround the gonads, the Leydig's organ within the walls of their esophagus, and a spiral valve in their intestine. These organs house typical immune cells (granulocytes, lymphocytes and plasma cells). They also possess an identifiable thymus and a well-developed spleen (their most important immune organ) where various lymphocytes, plasma cells and macrophages develop and are stored. Chondrostean fish (sturgeons, paddlefish and bichirs) possess a major site for the production of granulocytes within a mass that is associated with the meninges (membranes surrounding the central nervous system.) Their heart is frequently covered with tissue that contains lymphocytes, reticular cells and a small number of macrophages. The chondrostean kidney is an important hemopoietic organ; where erythrocytes, granulocytes, lymphocytes and macrophages develop.

Like chondrostean fish, the major immune tissues of bony fish (or teleostei) include the kidney (especially the anterior kidney), which houses many different immune cells. In addition, teleost fish possess a thymus, spleen and scattered immune areas within mucosal tissues (e.g. in the skin, gills, gut and gonads). Much like the mammalian immune system, teleost erythrocytes, neutrophils and granulocytes are believed to reside in the spleen whereas lymphocytes are the major cell type found in the thymus. In 2006, a lymphatic system similar to that in mammals was described in one species of teleost fish, the zebrafish. Although not confirmed as yet, this system presumably will be where naive (unstimulated) T cells accumulate while waiting to encounter an antigen.

Fish Diseases and Parasites

Like humans and other animals, fish suffer from diseases and parasites. Fish defences against disease are specific and non-specific. Non-specific defences include skin and scales, as well as the mucus layer secreted by the epidermis that traps microorganisms and inhibits their growth. If pathogens breach these defences, fish can develop inflammatory responses that increase the flow of blood to infected areas and deliver white blood cells that attempt to destroy the pathogens.

Specific defences are specialised responses to particular pathogens recognised by the fish's body, that is adaptative immune responses. In recent years, vaccines have become widely used in aquaculture and ornamental fish, for example vaccines for furunculosis in farmed salmon and koi herpes virus in koi.

Some commercially important fish diseases are VHS, ich and whirling disease.

Figure: *This gizzard shad has VHS, a deadly infectious disease which causes bleeding. It afflicts over 50 species of freshwater and marine fish in the northern hemisphere.*

Disease

All fish carry pathogens and parasites. Usually this is at some cost to the fish. If the cost is sufficiently high, then the impacts can be characterised as a disease. However disease in fish is not understood well. What is known about fish disease often relates to aquaria fish, and more recently, to farmed fish.

Disease is a prime agent affecting fish mortality, especially when fish are young. Fish can limit the impacts of pathogens and parasites with behavioural or biochemical means, and such fish have reproductive advantages. Interacting factors result in low grade infection becoming fatal diseases. In particular, things that causes stress, such as natural droughts or pollution or predators, can precipitate outbreak of disease.

Disease can also be particularly problematic when pathogens and parasites carried by introduced species affect native species. An introduced species may find invading easier if potential predators and competitors have been decimated by disease.

Pathogens which can cause fish diseases comprise:

- viral infections
- bacterial infections, such as *Pseudomonas fluorescens* leading to fin rot and fish dropsy
- fungal infections
- water mould infections, such as *Saprolegnia* sp.
- metazoan parasites, such as copepods
- unicellular parasites, such as *Ichthyophthirius multifiliis*
- Certain parasites like Helminths for example *Eustrongylides*

Parasites

Figure: *The isopod Anilocra gigantea parasitising the snapper Pristipomoides filamentosus*

Parasites in fish are a common natural occurrence. Parasites can provide information about host population ecology. In fisheries biology, for example, parasite communities can be used to distinguish distinct populations of the same fish species co-inhabiting a region.

Additionally, parasites possess a variety of specialized traits and life-history strategies that enable them to colonize hosts. Understanding these aspects of parasite ecology, of interest in their own right, can illuminate parasite-avoidance strategies employed by hosts.

Figure: Cymothoa exigua *is a parasitic crustacean which enters a fish through its gills and destroys the fish's tongue.*

Usually parasites (and pathogens) need to avoid killing their hosts, since extinct hosts can mean extinct parasites. Evolutionary constraints may operate so parasites avoid killing their hosts, or the natural variability in host defensive strategies may suffice to keep host populations viable. Parasite infections can impair the courtship dance of male threespine sticklebacks. When that happens, the females reject them, suggesting a strong mechanism for the selection of parasite resistance."

However not all parasites want to keep their hosts alive, and there are parasites with multistage life cycles who go to some trouble to kill their host. For example, some tapeworms make some fish behave in such a way that a predatory bird can catch it. The predatory bird is the next host for the parasite in the next stage of its life cycle. Specifically, the tapeworm *Schistocephalus solidus* turns infected threespine stickleback white, and then makes them more buoyant so that they splash along at the surface of the water, becoming easy to see and easy to catch for a passing bird.

Parasites can be internal (endoparasites) or external (ectoparasites). Some internal fish parasites are spectacular, such as the philometrid nematode *Philometra fasciati* which is parasitic in the ovary of female Blacktip grouper; the adult female parasite is a red worm which can reach up to 40 centimetres in length, for a diameter of only 1.6 millimetre; the males are tiny. Other internal parasites are found living inside fish gills, include encysted adult didymozoid trematodes, a few trichosomoidid nematodes of the genus *Huffmanela*, including *Huffmanela ossicola* which lives within the gill bone, and the encysted parasitic turbellarian *Paravortex*. Various protists and Myxosporea are also parasitic on gills, where they form cysts.

Fish gills are also the preferred habitat of many external parasites, attached to the gill but living out of it. The most common are monogeneans and certain groups of parasitic copepods, which can be extremely numerous. Other external parasites found on gills are leeches and, in seawater, larvae of gnathiid isopods. Isopod fish parasites are mostly external and feed on blood. The larvae of the Gnathiidae family and adult cymothoidids have piercing and sucking mouthparts and clawed limbs adapted for clinging onto their hosts. *Cymothoa exigua* is a parasite of the spotted rose snapper *Lutjanus guttatus*. It causes the tongue of the fish to atrophy and takes its place in what is believed to be the first instance discovered of a parasite functionally replacing a host structure in animals.

Other parasitic disorders, include *Gyrodactylus salaris*, *Ichthyophthirius multifiliis*, cryptocaryon, velvet disease, *Brooklynella hostilis*, Hole in the head, *Glugea*, *Ceratomyxa shasta*, *Kudoa thyrsites*, *Tetracapsuloides bryosalmonae*, *Cymothoa exigua*, leeches, nematode, flukes, carp lice and salmon lice.

Although parasites are generally considered to be harmful, the eradication of all parasites would not necessarily be beneficial. Parasites account for as much as or more than half of life's diversity; they perform an important ecological role (by weakening prey) that ecosystems would take some time to adapt to; and without parasites organisms may eventually tend to asexual reproduction, diminishing the diversity of sexually dimorphic traits. Parasites provide an opportunity for the transfer of genetic material between species. On rare, but significant, occasions this may facilitate evolutionary changes that would not otherwise occur, or that would otherwise take even longer.

Cleaner Fish

Some fish take advantage of cleaner fish for the removal of external parasites. The best known of these are the Bluestreak cleaner wrasses of the genus *Labroides* found on coral reefs in the Indian Ocean and Pacific Ocean. These small fish maintain so-called "cleaning stations" where other fish, known as hosts, will congregate and perform specific movements to attract the attention of the cleaner fish. Cleaning behaviours have been observed in a number of other fish groups, including an interesting case between two cichlids of the same genus, *Etroplus maculatus*, the cleaner fish, and the much larger *Etroplus suratensis*, the host.

Figure: *Two cleaner wrasses, Labroides phthirophagus, servicing a goatfish, Mulloidichthys flavolineatus*

More than 40 species of parasites may reside on the skin and internally of the ocean sunfish, motivating the fish to seek relief in a number of ways. In temperate regions, drifting kelp fields harbour cleaner wrasses and other fish which remove parasites from the skin of visiting sunfish.

In the tropics, the *mola* will solicit cleaner help from reef fishes. By basking on its side at the surface, the sunfish also allows seabirds to feed on parasites from their skin. Sunfish have been reported to breach more than ten feet above the surface, possibly as another effort to dislodge parasites on the body.

Mass Die Offs

Some diseases result in mass die offs. One of the more bizarre and recently discovered diseases produces huge fish kills in shallow marine waters. It is caused by the ambush predator dinoflagellate *Pfiesteria piscicida*. When large numbers of fish, like shoaling forage fish, are in confined situations such as shallow bays, the excretions from the fish encourage this dinoflagellate, which is not normally toxic, to produce free-swimming zoospores. If the fish remain in the area, continuing to provide nourishment, then the zoospores start secreting a neurotoxin. This toxin results in the fish developing bleeding lesions, and their skin flakes off in the water.

The dinoflagellates then eat the blood and flakes of tissue while the affected fish die. Fish kills by this dinoflagellate are common, and they may also have been responsible for kills in the past which were thought to have had other causes. Kills like these can be viewed as natural mechanisms for regulating the population of exceptionally abundant fish. The rate at which the kills occur increases as organically polluted land runoff increases.

Wild Salmon

Figure: Henneguya salminicola, *a protozoan parasite*

According to Canadian biologist Dorothy Kieser, protozoan parasite *Henneguya salminicola* is commonly found in the flesh of salmonids. It has been recorded in the field samples of salmon returning to the Queen Charlotte Islands. The fish responds by walling off the parasitic infection into a number of cysts that contain milky fluid. This fluid is an accumulation of a large number of parasites.

Henneguya and other parasites in the myxosporean group have a complex lifecycle where the salmon is one of two hosts. The fish releases the spores after spawning. In the *Henneguya* case, the spores enter a second host, most likely an invertebrate, in the spawning stream. When juvenile salmon out-migrate to the Pacific Ocean, the second host releases a stage infective to salmon.

The parasite is then carried in the salmon until the next spawning cycle. The myxosporean parasite that causes whirling disease in trout, has a similar lifecycle. However, as opposed to whirling disease, the *Henneguya* infestation does not appear to cause disease in the host salmon — even heavily infected fish tend to return to spawn successfully.

According to Dr. Kieser, a lot of work on *Henneguya salminicola* was done by scientists at the Pacific Biological Station in Nanaimo in the mid-1980s, in particular, an overview report which states that "the fish that have the longest fresh water residence time as juveniles have the most noticeable infections. Hence in order of prevalence coho are most infected followed by sockeye, chinook, chum and pink." As well, the report says that, at the time the studies were conducted, stocks from the middle and upper reaches of large river systems in British Columbia such as Fraser, Skeena, Nass and from mainland coastal streams in the southern half of B.C. "are more likely to have a low prevalence of infection." The report also states "It should be stressed that *Henneguya*, economically deleterious though it is harmless from the view of public health. It is strictly a fish parasite that cannot live in or affect warm blooded animals, including man".

According to Klaus Schallie, Molluscan Shellfish Programme Specialist with the Canadian Food Inspection Agency, "*Henneguya salminicola* is found in southern B.C. also and in all species of salmon. I have previously examined smoked chum salmon sides that were riddled with cysts and some sockeye runs in Barkley Sound (southern B.C., west coast of Vancouver Island) are noted for their high incidence of infestation."

Sea lice, particularly *Lepeophtheirus salmonis* and a variety of *Caligus* species, including *Caligus clemensi* and *Caligus rogercresseyi*, can cause deadly infestations of both farm-grown and wild salmon. Sea lice are ectoparasites which feed on mucous, blood, and skin, and migrate and latch onto the skin of wild salmon during free-swimming, planktonic *naupli* and *copepodid* larval stages, which can persist for several days. Large numbers of highly populated, open-net salmon farms can create exceptionally large concentrations of sea lice; when exposed in river estuaries containing large numbers of open-net farms, many young wild salmon are infected, and do not survive as a result. Adult salmon may survive otherwise critical numbers of sea lice, but small, thin-skinned juvenile salmon migrating to sea are highly vulnerable. On the Pacific coast of Canada, the louse-induced mortality of pink salmon in some regions is commonly over 80%.

Farmed Salmon

Figure: *Atlantic salmon*

In 1972, Gyrodactylus salaris, also called salmon fluke, a monogenean parasite, spread from Norwegian hatcheries to wild salmon, and devastated some wild salmon populations.

In 1984, infectious salmon anemia (ISAv) was discovered in Norway in an Atlantic salmon hatchery. Eighty percent of the fish in the outbreak died. ISAv, a viral disease, is now a major threat to the viability of Atlantic salmon farming.

It is now the first of the diseases classified on List One of the European Commission's fish health regime. Amongst other measures, this requires the total eradication of the entire fish stock should an outbreak of the disease be confirmed on any farm. ISAv seriously affects salmon farms in Chile, Norway, Scotland and Canada, causing major economic losses to infected farms.

As the name implies, it causes severe anemia of infected fish. Unlike mammals, the red blood cells of fish have DNA, and can become infected with viruses. The fish develop pale gills, and may swim close to the water surface, gulping for air. However, the disease can also develop without the fish showing any external signs of illness, the fish maintain a normal appetite, and then they suddenly die. The disease can progress slowly throughout an infected farm and, in the worst cases, death rates may approach 100 percent. It is also a threat to the dwindling stocks of wild salmon. Management strategies include developing a vaccine and improving genetic resistance to the disease.

In the wild, diseases and parasites are normally at low levels, and kept in check by natural predation on weakened individuals. In crowded net pens they can become epidemics. Diseases and parasites also transfer from farmed to wild salmon populations. A recent study in British Columbia links the spread of parasitic sea lice from river salmon farms to wild pink salmon in the same river.

The European Commission (2002) concluded :

> *"The reduction of wild salmonid abundance is also linked to other factors but there is more and more scientific evidence establishing a direct link between the number of lice-infested wild fish and the presence of cages in the same estuary."*

It is reported that wild salmon on the west coast of Canada are being driven to extinction by sea lice from nearby salmon farms. Antibiotics and pesticides are often used to control the diseases and parasites.

Coral Reef Fish

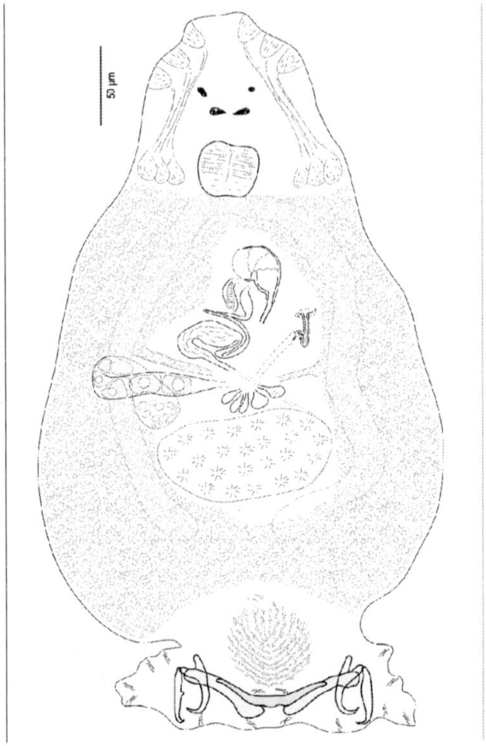

Figure: *A monogenean parasitic on the Blacktip grouper*

Coral reef fish are characterized by high biodiversity. As a consequence parasites of coral reef fish show tremendous variety. Parasites of coral reef fish include nematodes, Platyhelminthes (cestodes, digeneans, and monogeneans), leeches, parasitic crustaceans such as isopods and copepods, and various microorganisms such as myxosporidia and microsporidia. Some of these fish parasites have heteroxenous life cycles (i.e. they have several hosts) among which sharks (certain cestodes) or molluscs (digeneans). The high biodiversity of coral reefs increases the complexity of the interactions between parasites and their various and numerous hosts. Numerical estimates of parasite biodiversity have shown that certain coral fish species have up to 30 species of parasites. The mean number of parasites per fish species is about ten. This has a consequence in term of co-extinction. Results obtained for the coral reef fish of New Caledonia suggest that extinction of a coral reef fish species of average size would eventually result in the co-extinction of at least ten species of parasites.

Bibliography

Allan, J.D.: *Stream Ecology: structure and function of running waters*: Chapman and Hall, London, 1995.

Blatteis, Clark M.: *Physiology and Pathophysiology of Temperature Regulation*, World Scientific Publishing Co., NJ, 2001.

Bone, Q; Moore, R H: *Biology of Fishes*, Taylor & Francis Group, United Kingdom, 2008.

Cushing, C.E. and J.D. Allan: *Streams: their ecology and life*, Academic Press, San Diego, 2001.

Farrell, Anthony P.: *Encyclopedia of Fish Physiology: From Genome to Environment*, Academic Press, United States, 2008.

Fraser, L.H. and P.A. Keddy: *The World's Largest Wetlands: Ecology and Conservation*, Cambridge University Press, Cambridge, UK., 2005.

Gilbert, Scott F.: *Developmental Biology*, Sinauer Associates, Sunderland, Massachusetts, 1994.

Giller, S. and B. Malmqvist: *The Biology of Streams and Rivers*, Oxford University Press, Oxford, 1998.

Goulding, M.: *The Fishes and the Forest:Explorations in Amazonian Natural History*, University of California Press, Berkeley, CA, 1980.

Helfman, G.; Collette; Facey, D.; Bowen, BW: *The Diversity of Fishes: Biology, Evolution, and Ecology*, Wiley-Blackwell, Hoboken, New Jersey, 2009.

Hoyle G, Wiersma CAG: *Identified neurons and behavior of arthropods*, Plenum Press, New York, 1977.

Kardong, K.: *Vertebrates: Comparative anatomy, function, evolution*, McGraw-Hill, Boston, 2008.

Keddy, P.A.: *Wetland Ecology: Principles and Conservation*, Cambridge University Press, Cambridge, UK., 2010.

Keddy, Paul A.: *Wetland ecology : principles and conservation*, Cambridge University Press, New York, 2010.

Kotpal, R. L.: *Modern Text Book of Zoology*: *Vertebrates*, Rastogi Publications, Shivaji Road Meerut, U.P., India, 2010.

Liem, KF: *In Environmental Physiology of Fishes,* Plenum Press, New York, 1980.

McGinnis, Samuel M.: *Field Guide to Freshwater Fishes of California*, University of California Press, 2006.

Mitsch, W.J., J.G. Gosselink, C.J. Anderson, and L. Zhang, *Wetland Ecosystems*, John Wiley & Sons, New York, 2009

Mitsch, William J.; James G. Gosselink: *Wetlands*, John Wiley & Sons, New York, 2007.

Moyle, PB; Cech, J. J.: *Fishes, An Introduction to Ichthyology,* Benjamin Cummings, United States, 2003.

Mungall C. and McLaren, D.J.: *Planet under stress: the challenge of global change*, Oxford University Press, New York, USA, 1991.

Nelson, Joseph S.: *Fishes of the World*, John Wiley & Sons, United States, 2006

Ostrander, Gary Kent: *The Laboratory Fish*, Elsevier, Amsterdam, Netherlands, 2000.

Paxton, J.R. & Eschmeyer, W.N.: *Encyclopedia of Fishes*, Academic Press, San Diego, 1998.

Piper, Ross, *Extraordinary Animals: An Encyclopedia of Curious and Unusual Animals*, Greenwood Press, United States, 2007.

Prosser, C. Ladd: *Comparative Animal Physiology, Environmental and Metabolic Animal Physiology,* Wiley-Liss, Hoboken, NJ, 1991.

Pullin, Andrew S.: *Conservation biology*, Cambridge University Press, New York, 2002.

Ricciuti, Edward R.; Bird, Jonathan: *Killers of the Seas: The Dangerous Creatures That Threaten Man in an Alien Environment*, The Lyons Press Guilford, CT., 2003.

Robert, G. Wetzel and Gene, E. Likens: *Limnological Analyses*, Springer-Verlag, New York City, 2000)

Schindler, David and Vallentyne, John R.: *Over fertilization of the World's Freshwaters and Estuaries*, University of Alberta Press, 2004.

Shiklomanov, I A,: *World fresh water resources in Glick*, Oxford University Press, United Kingdom, 1993.

Simmons PJ, Young D.: *Nerve cells and animal behaviour*, Cambridge University Press, London, 1999.

Stein PSG: *Neurons, Networks, and Motor Behavior*, MIT Press, United States, 1999.

Vara, A.; Zapata, A.G. and Chiba, A.: *Cells and tissues of the immune system of fish*, Academic Press, New York, 2008.

WC, Aird: *Endothelial biomedicine*, Cambridge University Press, London, 2007.

Index

❑❑❑